This is an intriguing and challenging book. Its contributors, characters, and constituents seize upon and severally straddle the terms, textures, and tangles of long-ranging pasts and diverse practices in South Asia. Taken together, the explorations on offer of 'vernacular mobilities' do more than what meets the eye. They underscore instead the urgent requirements today of rethinking and unthinking not only 'coloniality' and 'postcoloniality' but equally 'decoloniality' and 'travel'. And to do so by tracking the enticements and incitements of these formidable fabulations – as metaphor, optic, and provocation.

– **Saurabh Dube**, *Distinguished Professor, El Colegio de México, Mexico City & Distinguished Research Fellow, Max Weber Forum for South Asian Studies, Germany-India*

Departing wisely from the long-cherished prism of 'precolonial-religious and post-colonial-secular', this important book builds upon recent work in the field of colonial and vernacular modernities project in South Asia. Drawing from the rich literary landscape of the Indian subcontinent, the contributors explore variously 'alternative travel performances', in spatial and temporal terms, embracing a vast and rich cornucopia of travel narratives. This is a pioneering volume: original, insightful, and provocative. A valuable contribution to Indian Travel Writings and South Asia Studies.

– **Sachidananda Mohanty,** *Editor of* Travel Writings and the Empire, *Former Professor of English, University of Hyderabad & Former Vice-Chancellor, Central University of Odisha*

Decolonial Travel

This volume brings together scholarship on indigenous forms of travel to decolonize travel theory. It looks at certain minoritarian-vernacular travelling cults – very rarely examined – that compel us to rethink, on the one hand, the conventional tropes of and rationales for travel, and, on the other hand, notions of (post)coloniality, nationalism and modernity in the context of India. The book illustrates the enduring problematic of the 'colonial episteme': how it deploys pervasive categories through which travel practices are sought to be understood, and why such categories are inadequate in accounting for the vernacular travelling cults in question. In studying the vernacular world-making in and through these cults, this book offers critical insights on how they defy the log(ist)ics of the 'imperial categories' and why they must be read as expressions of decoloniality. An important contribution to travel studies, the book will be an indispensable resource for students and researchers of South Asian studies, travel theory, Indian literary and cultural studies, cultural history and anthropology, sociology and decoloniality.

Avishek Ray teaches at the National Institute of Technology Silchar. He is the author of *The Vagabond in the South Asian Imagination: Representation, Agency & Resilience* (2021) and co-editor of *Nation, Nationalism and the Public Sphere: Religious Politics in India* (2020). His research interests span space and mobility, (post)nationalism and postcolonialism. In 2021, he was awarded a Fulbright-Nehru Academic and Professional Excellence Fellowship.

Decolonial Travel

Vernacular Mobilities in India

**Edited by
Avishek Ray**

Routledge
Taylor & Francis Group

LONDON AND NEW YORK

First published 2025
by Routledge
4 Park Square, Milton Park, Abingdon, Oxon OX14 4RN

and by Routledge
605 Third Avenue, New York, NY 10158

Routledge is an imprint of the Taylor & Francis Group, an informa business

British Library Cataloguing-in-Publication Data
A catalogue record for this book is available from the British Library

ISBN: 978-1-032-85808-1 (hbk)
ISBN: 978-1-032-89770-7 (pbk)
ISBN: 978-1-003-54452-4 (ebk)

DOI: 10.4324/9781003544524

Typeset in Times New Roman
by Deanta Global Publishing Services, Chennai, India

Contents

Acknowledgements   *vii*
Author Biographies   *viii*

Introduction   1

1   Decolonizing Travel(ing Theory), or the Discursive Limits of the '(Post)colonial'   3
AVISHEK RAY

The Pre-Colonial and the Scriptural   13

2   To go is to Know (That You Never Went): A Sanskrit Buddhist Map of Selected Illusions   15
MATTIA SALVINI

3   Vedic Travel: The Agnihotra and Beyond   30
LAUREN M. BAUSCH

The State, Polity and the Religious   77

4   Travel(ing) to Write: Authorship and Agency in an Era of Inter-polity Mobility   79
RAFIA KHAN

5   The Power of Itinerancy: Religious Leaders in the Nepal-India Borderland   93
MARTIN GAENSZLE

Language and the Literary 107

6   'Floating Straight Obedient to the Stream': Bibliomigrancy
    and Riverine Journeys in Colonial Bengal   109
    SWATI CHATTOPADHYAY

7   Moving in Circles: *Chakkars* in Rajasthani Women's Songs   127
    NILANJANA MUKHERJEE AND GAURAV KUMAR

The Trope of Homelessness 143

8   Homeless in Gujarat and India: On the Curious Love of
    Indulal Yagnik   145
    AJAY SKARIA

9   The Homeless Gandhi   171
    VINAY LAL

    Index   179

Acknowledgements

Editing this book has been very instructive. Working on book projects always forges new collaborations. I have learnt a great deal in the process of collaboration with the contributors. I thank them for this rewarding experience.

In the summer of 2021, I convened a symposium 'Decolonizing Travel(ing) Theory Explorations in the "Indian" Analytic Traditions'. Issues that are presented in this book were first discussed here. The Indian Council of Philosophical Research (ICPR) funded the symposium. I thank ICPR for the funds and, by extension, making this book possible. Among the contributors, Mattia Salvini, Lauren Bausch, and Swati Chattopadhyay presented preliminary drafts of their chapters at the symposium. Other speakers for the symposium included Saurabh Dube, Charu Gupta, Maya Joshi, and David Ludden. Although they could not write for this volume, their critical insights on the theme – those that they brought at the symposium – helped in thinking through this project. I thank them profusely for this.

The Introduction to this volume is a slightly reworked version of my article, 'Decolonizing Travel(ing) Theory: Vernacular Travels in the Indian Context', published in Issue 124 of *Cultural Critique* (Summer 2024), pp. 164–176. I thank the journal and University of Minnesota Press for allowing me to reproduce the article in this volume. Chapter 8 of the volume has also eponymously appeared in *The Indian Economic & Social History Review*, Vol. 38(3), pp. 271–297. I thank Ajay Skaria, the journal, and SAGE for allowing me to reprint it here.

My sincere gratitude goes to Giuliano Giustarini, Gaurab Chattopadhyay, Sudeshna Datta Chaudhuri, Debaditya Mukhopadhyay, and Rafia Khan for generously helping me review certain chapters for this volume, and to Agnidh Baruah, Suhash Bhattacharjee and Diyali Bhattacharya for formatting and indexing the manuscript. Finally, I would like to thank Routledge for commissioning this project and rendering their editorial and professional support during its preparation.

Author Biographies

Ajay Skaria is a scholar of South Asian Politics and History and is associated with Postcolonial and Subaltern Studies. He is currently teaching at the University of Minnesota in the Department of History. He is the author of *Hybrid Histories: Forests, Frontiers and Wildness in Western India* (Oxford University Press).

Gaurav Kumar is Assistant Professor of English at Shaheed Bhagat Singh College, University of Delhi. He recently completed his tenure as a UGC Junior Research Fellow in English Literature, pursuing an MPhil at the Department of Comparative Literature and Translation Studies, Ambedkar University, Delhi. He holds both a bachelor's and a master's degree in English Literature from the University of Delhi. His MPhil dissertation examines the philosophy of desire, laughter, and translation in minor literatures of India. His areas of interest include comparative critical and literary theory.

Lauren M. Bausch is Professor at Dharma Realm Buddhist University, where she teaches Sanskrit, Indian Classics, and Buddhism. A specialist in the Brāhmaṇa texts, she also investigates the relationship between Vedic tradition and early Indian Buddhism. She is currently writing a book on Vedic philosophy in the Brāhmaṇa texts. She is the editor of *Self, Sacrifice, and Cosmos: Vedic Thought, Ritual, and Philosophy* (2019) and the author of articles such as 'The Kāṇva Brāhmaṇas and Buddhists in Kosala', 'Philosophy of Language in the Ṛgveda', and 'Dual Varṇa in Vedic Texts'. She completed her PhD in Sanskrit in the Department of South and Southeast Asian Studies at the University of California, Berkeley, in 2015.

Martin Gaenszle is Professor (retired) in Cultural and Intellectual History of Modern South Asia at the University of Vienna, Austria. His scholarly interests include religious pluralism, ethnicity, local history, and oral traditions in South Asia, particularly in the Himalayan region. He has done field research in Eastern Nepal and North India and published books on Rai oral traditions, ritual speech, and concepts of space in Banaras. More recently, he has worked on ethno-religious reform movements in Nepal.

Mattia Salvini Dharmavardhana Jñānagarbha studied Sanskrit and Indian Philosophy mostly in India. He obtained a BA and an MA in Sanskrit from RKM Vivekananda College (Madras University), India, and a PhD from the

SOAS University of London. He has published articles and book chapters on Madhyamaka philosophy with the *Journal of Indian Philosophy, Thai International Journal of Buddhist Studies, the Bulletin de l'École française d'Extrême-Orient*, Oxford University Press, and Routledge. Presently, Mattia is the Rector and the Dean of the Faculty of Graduate Studies at the International Buddhist College, Sadao Campus, Thailand, and continues to research on Buddhist philosophical texts and other aspects of Buddhism in India.

Nilanjana Mukherjee teaches English Literature at Shaheed Bhagat Singh College, University of Delhi. She completed her PhD in the year 2011 from Jawaharlal Nehru University, New Delhi. She is the author of the book, *Spatial Imaginings in the Age of Colonial Cartographic Reason: Maps, Landscapes, Travelogues in Britain and India* (Routledge: 2020). She has also co-edited the volume, *Mapping India: Transitions and Transformations 18th-19th Centuries* (Routledge: 2019). In the past, she was the Charles Wallace India Trust Visiting Fellow at the Institute of Advanced Studies in the Humanities, University of Edinburgh, in the year 2017 and a doctoral fellow at King's College, London, in 2009. At present, she is working on a study of the Indian Desert as a spatial imaginary in its evolution as a frontier zone in Northwest India.

Rafia Khan is Assistant professor (Liberal Arts) at Nirma University, Ahmedabad, India. She was awarded her doctorate from the Centre for Historical Studies, JNU, New Delhi in 2021. Her interest lies in studying socio-political formations; exploring material-cultural and socio-political aspects of the expression and articulation of political authority; and the shaping of individual, group, and communal identities at the regional, local, and sub-local levels in the twelfth-fifteenth centuries.

Swati Chattopadhyay, Professor in the Department of History of Art and Architecture at the University of California, Santa Barbara, is an architectural and urban historian, specializing in modern architecture and the cultural landscape of the British Empire. She is the author of *Representing Calcutta: Modernity, Nationalism, and the Colonial Uncanny* (Routledge, 2005), *Unlearning the City: Infrastructure in a New Optical Field* (Minnesota, 2012), and *Small Spaces: Recasting the Architecture of Empire* (Bloomsbury, 2023).

Vinay Lal is a historian of India. He is Professor of History and Asian American Studies at the University of California, Los Angeles. He writes widely on the history and culture of colonial and modern India, popular and public culture in India, cinema, historiography, the politics of world history, the Indian diaspora, global politics, contemporary American politics, the life and thought of Mohandas Gandhi, Hinduism, and the politics of knowledge systems.

Introduction

1 Decolonizing Travel(ing Theory), or the Discursive Limits of the '(Post)colonial'[1]

Avishek Ray

Travelling, both as a concept and performativity, has yielded a diverse range of critical literatures that probe into the epistemological, political, ethical, and aesthetic dimensions of mobility. However, much of the critical theories on travel (and, of late, nomadology) draw on the Western canon, while there appears to be a dearth of proportionate research on and/or documentation of the decolonial analytical frameworks that engage with vernacular travel and theories thereof, what I refer to as 'travel(ing) theory'. Referencing Edward Said (1983), this volume uses 'travel(ing) theory' as a generative concept to better understand the vernacular travelling cults in the 'local' contexts that 'foreign' theories – often a-historically and a-culturally adapted in and adopted by 'local' contexts – cannot make sense of. So, here, we begin with the proposition that certain ways of vernacular travelling are not amenable to the array of theorizations we have at our disposal. In this volume, we examine the limitations of the theoretical frameworks that are generally brought to bear on studying travel practices and the genre of travel writing, particularly with reference to the postcolonial Indian context.

Many scholars have approached (post)colonial travel (both as an embodied performance and as textual practice) through conceptual/analytical categories already implicated in a dichotomous colonial framework – e.g., 'pre-modern religious' versus 'modern secular', 'colonial' versus 'postcolonial', all categories invariably laden with value judgements. Breaking away from such an approach, this volume draws on the critical scholarship on decoloniality. It proposes an alternative theoretical framework that situates the decolonial travel(ler)s beyond the prevailing postcolonial regime: readings and renderings that, sometimes rather uncritically, deploy the categories 'pre-modern religious' and 'modern secular' as staple. Indeed, 'pre-modern' India has had a strong tradition of travelling, alongside producing an extensive body of works exploring the ethics and politics of travelling. All Indic orders of monkhood – Hindu (cf. *Sannyasa Upanishad*), Buddhist (cf. *Samannaphalasutta*), or Jain – for instance, command travelling to be the chief 'marker' of the renunciant, thus distinguished from the householder or the 'laity'. Concepts of the *atithi* (stranger who warrants unconditional hospitality), that of *paribrajaka* (wandering mendicant), and genres like the *seyahatnâme*or (the practice of positing the *Sufi-Bhakti* figures as prophetic 'mad nomads') are abound in the 'pre-modern' Indian cultural repertoire, but least theorized from the point of view of their vernacularity. My bid is to suggest that these long-standing vernacular

DOI: 10.4324/9781003544524-2

travelling cults did not leave behind any epistemological framework with which one can make sense of their vernacularity.

Glimpsed in religious texts, archaeological objects, and oral traditions, these practices of travel have been examined primarily as 'historical evidence' read through an evolutionary-teleological narrative of the 'pre-modern' (with which imageries of the 'modern secular' are contrasted). How else might they be understood? What does it mean to view them through a non-teleological frame? What specific forms of embodied 'mobility' do these travels illustrate? What new conceptual categories do we need to invoke? What notions of indigeneity – both with reference to travelling practices and theorizations thereof – do they invoke? What conceptual vocabularies do they endow us with? And, why are they not amenable to existing theorizations?

The essays in this volume do not necessarily answer these questions, but seek to furnish a preliminary but provocative framework for decolonial thinking, which is at best a 'gesture' toward alternative imageries of vernacular travel. In this volume, we set the term 'mobility' (of which 'travel' is a subset) in motion around the vernacular travel cults. The recognition of travel as 'travel', that is, its canonization, is reliant upon, first, a-religious documentation, and second, the agency of the 'modern' subject with access to 'leisure' as a corollary of the Protestant work ethic. Thinking in these terms, ideas of travel are already steeped in the Western-imperialist epistemology. This volume proposes a theoretical preamble to decouple the decolonial practices of mobility in India from such underpinnings. And, in so doing, it offers a conceptual framework in which decolonial mobility signals a flexible mode of theorization – as indicated in the notion of 'traveling theory' – that might be intellectually dexterous and tractable in the postcolonial Indian context, rather than relying on fixed taxonomies and/or binaries.

However, the principal difficulty in making any forays in this direction is the lack of 'historical' documentation of the vernacular travelling cults. Historical documentations, we imply, use a chronological/teleological (rather than religious) analytic guided by the Hegelian (2007) vision of objectivist history-writing, supremely guided by Reason (with a capital 'R'). This practice would be ushered by the formalization of history as a discipline and subsequently emerge as a corollary of the Enlightenment project.[2] It is then perhaps anachronistic to seek 'historical' documentations of travel in the 'pre-modern' genres of expression. Indeed, these genres, such as religious scriptures, literary texts, archaeological artefacts, etc., evidence travelling. However, they serve as 'outside sources' – sources not intrinsic to travelling – from where historical evidence on travelling may be gleaned, but they are not per se 'history' in the Hegelian sense of the term. For example, the genre of the *Safarnama* of Arabic-Persian origin, or the genre of quasi-fictional *Samdesa Kavya* in the Sanskritic tradition, is a descriptive narrative of travel. One can read it as a historical resource, but it is not 'history' as such. According to Benjamin Walker (1938, p. 520), '[t]he ancient Indians seem to have been great travellers, although they preserved no record of their wanderings, and information is to be gleaned only from outside sources, or by inference'. Needless to say, inferences entail interpretation, subjective and heuristically-contested. Like all interpretations, these are then to be read as narratives, socially constructed, ideologically mediated, and maneuvered in order to meet the requirements of changing historical periods.

This problem is amplified in the (post)colonial era. The 'colonial episteme' (Cohn, 1996) could not make sense of certain vernacular practices of mobility when approached from the perspective of instrumental rationality and viewed through 'imperial eyes' (Pratt, 1992). Distinguishing the 'good traveler' from the 'bad traveler' became a colonial agenda, and within the remit of this taxonomic practice, a whole series of new heuristic categories such as 'vagabond', 'vagrant', 'miscreants', 'nomadic criminal tribe', etc. would be invoked (drawing on the imperialist vocabulary) and deployed as a catch-all to indiscriminately cast out travelers and itinerants who posed an enigma for the colonial episteme. These categories, one must recognize, are social and historical constructs, rather than any universal states of being (Ray, 2020, 2021). In this context, Sanjay Nigam (1990a, 1990b) and Shail Mayaram (2003, pp. 13–35) discuss how the colonial state segregated the 'bad traveler' as a potential criminal, using quasi-scientific analytics of race, genetics, anthropometry, etc. However, the anxiety around the heuristics of travel continued to reflect in certain vernacular travelling cults as illustrated by 'revolutionaries' like Rahul Sankrityayan (1893–1963), Indulal Yagnik (1892–1972), Gandhi (1869–1948), Agyeya (1911–1987), among others.[3] In their postcolonial rebuttal, these figures continued to consciously wear the badge of 'bad traveler'. Consider, in this context, how Gandhi, for example, would travel by rail in his customary indigenous sartorial style – quite complacently so because his '"familiarity with the minor amenities of Western civilization" has taught me [him] to respect my [his] national costume' (Gandhi, 1964, p. 450) – without caring about the fact that this would, in the colonial gaze, provoke a 'nauseating and humiliating spectacle of Gandhiji striding half naked' (Churchill, cited in Fischer, 1962, p. 281). It is important here to observe that what Gandhi refers to as 'national costume' is, in fact, far from any 'national costume', his inimitably self-fashioned sartorial style, and not something the average Indian would wear. Gandhi (1964, p. 450) himself admitted that his 'countrymen…[were into] wearing…semi-European dress in the courts and elsewhere'. Clearly, by 'national costume', he meant 'indigenous attire'. What is interesting, however, in the course of our discussion, is the way in which 'national' and 'indigenous' are swapping places at the (post)colonial juncture. I use Gandhi's example here to underscore the nature of epistemic transformation wrought upon by colonial governmentality, and how certain 'nomad natives' thought about (counter-)mobility in response. We have to be, therefore, attentive to the wider ramifications these seemingly innocuous and quotidian acts of mobility may have for the configurations of 'Indian' (post)coloniality, nationalism, and modernity; and how the 'nomad natives' themselves embody 'the myriad ways ordinary people guide, deceive, and toy with power instead of confronting it directly' (Mbembe, 2001, p. 128).

Although these cults are unanimously recognized as a significant corpus of local knowledge production and representation, the repertoire of postcolonial scholarship characterizes a historicist-textualist tendency to posit these cults as a springboard to 'write back to the empire'. It ascribes a 'historical function' to these travel(ogue)s.

The vernacular travel cults then become indexical to the *zeitgeist*, a window to the postcolonial consciousness, a medium and method for reversing the colonial gaze. In this scheme, travel writings become a means-to-an-end (the end being historical inferences), but, strictly speaking, travel-in-itself has little 'worth'. This entails logocentrism: here, travel *writing* overrides the phenomenological aspect of travel. I have elsewhere demonstrated how vernacular travel performance – the very act of mobility, besides textual performance – may embody new conceptual imageries of travel (Ray, 2021). However, within the remit of postcolonial scholarship, vernacular travel cults, when approached from a *longue durée* perspective, become subservient to the *telos* of postcoloniality; they are dovetailed to fit the postcolonial historiographic template (of India-versus-the-West), but the *act* of travel is not necessarily studied with the intent of unpacking the new epistemic categories, alternative imaginaries and the novel subjectivities – for instance, *ghumakkari* (cf. Rahul Sankrityayan), *jivanmukti* (cf. Rabindranath Tagore) and (riverine) *parikrama* (cf. Amritlal Vegad) – that are already embedded in the cults.

While some scholars are hard-pressed for evidence of 'leisure travel' – secular travel undertaken with formative goals – in the pre-modern Indian cultural repertoire (Alam & Subramaniam, 2007; Sen, 2005), others cite pilgrimage as an example of 'connected histories' and networks within and beyond 'pre-modern' India (Gupta, 2017, p. 316; Eck, 2012). Accordingly, postcolonial scholars – for example, Partha Chatterjee (1998), Kumkum Chatterjee (1999), Bhaskar Mukhopadhyay (2002), and Simonti Sen (2005), among others – institute a clear distinction between the 'pre-modern-religious' and the 'modern-secular' practices of travel. In brief, this dichotomy has emerged as the *only* heuristic outfit with reference to which scholars now think about vernacular travel. Even while recognizing the significance of the 'affordances' of the colonial milieu and the attendant 'modern' imperial infrastructures in bringing about a structural and attitudinal shift in travel practices, I contend that this framework of pre-modern-religious versus modern-secular, though only provisionally instructive, has not always been deployed with the discretionary judgement and utmost caution that it warrants. Sen's (2005) study of 'modern' Bengali travelogues, for example, is symptomatic of this problem. For this, I have taken her to task (2019). The problem is not so much with using this framework as with the pervasive tendency to unreflectively project it onto certain alternative travel performances that do not fit into the taxonomy. Such a projection then fails to appraise, as mentioned earlier, the alternative imaginaries and conditions that made possible the emergence of a new decolonial episteme of travelling that occasioned the articulation of vernacular modernity.

First and foremost, the 'epistemic rupture' between the 'pre-modern-religious' and the 'modern-secular', like all taxonomies, is premised on a classificatory practice that entails exclusionary politics and, therefore, perpetrates 'epistemic violence' (Foucault, 1972; Hacking, 1999). In parallel, it forges an imaginary coupling, on the one hand, between the 'pre-modern' and the 'religious', and, on the other hand, between the 'modern' and the 'secular'. However, these categories, at least in the curious case of South Asia, are not always neatly segregable; they are rather jumbled and uncannily 'hybrid' (Bhabha, 1994; Ray & Banerjee-Dube, 2020). In this context, Chatterjee (1997, p. 19) argues:

> [W]e construct a picture of 'those days' when there was beauty, prosperity and healthy sociability. This makes the very modality of our coping with modernity radically different from the historically evolved modes of Western modernity.

Drawing on Chatterjee, I argue that the segregable structure of the dichotomous framework – pre-modern/modern, religious/secular – fails to adequately account for the entire gamut of the vernacular sensibilities around the discourses of travel and how it lives out, negotiates, and contests 'the historically evolved modes of Western modernity'. Here, I have in mind the specific gestures of 'hybrid' travel rendered possible by the postcolonial milieu, following which vernacular travel practices started to meld uneasily with the modalities of 'modern', colonially endorsed, predominantly instrumentalist travel (Banerjee & Sandeep, 2015; Gupta, 2020; Nijhawan, 2020), and thereby, uneasily straddling the domains of the 'sublime-spiritual' and the 'secular', and that of the 'modern' and the 'unmodern'. Consider how the 'imperial order' imposed by the railways, for example, would help achieve a 'modern', capitalist spatio-temporal reorganization (Kerr, 2007; Ahuja, 2009), while at the same time intervening with the existing structures of vernacular mobility, leading to two apparently contrary brands of travelling practices – the 'modern' and the 'unmodern', the colonial and the counter-colonial, the sublime and the instrumentalist – bleeding into each other.

What has been referred to as 'vernacular travel' in the content of this volume emerges from South Asia's nativist response to the colonial encounter. Yet, I argue that pigeonholing these travels into the pre-existent categories of '(pre-)modern', '(non-)secular', or 'Indian' is tantamount to perpetrating what Miranda Fricker (2007, p. 1) calls 'hermeneutical injustice',[4] what is faced by a subject of knowledge when 'collective interpretive resources puts someone at an unfair disadvantage when it comes to making sense of their social experiences'. This approach obfuscates the myriad range of vernacular travel experiences – both indigenous and ingenious – and casts them into proleptic theoretical templates. Departing from here, this volume draws on Said's (1983) notion of 'traveling theory' and highlights the problem of conceptual/categorical (in)equivalencies among the three epistemes – the colonial, the postcolonial, and the decolonial – that have heavily influenced and mirrored each other. The transfer, reception and reappropriation of loan-concepts among the epistemes have led to interferences in the utilization of 'foreign' concepts in 'local' contexts. Paraphrasing Said (1983), one may call this phenomenon 'traveling concepts', whereby concepts from one episteme, though mappable with equivalent terminologies from another, may still signify different 'value', or alternatively, even without any terminological equivalence, may invoke identical signification in another episteme. Along these lines, we ask: is *ghumakkari* then the same as wandering, or likewise, *parikrama* the same as circumambulation? Are there exact terminologies/concepts for these expressions in the (post)colonial episteme, if it were to analyze them as 'subject of knowledge'? Is the 'essence' contained in these expressions and their performativity, notwithstanding the issue of conceptual (in)equivalencies, universalizable? Put differently, can it travel across epistemes?

In order to better understand these concerns, this volume gestures toward a methodological break from beyond the teleological-historicist lens of the dichotomous framework. It calls for new interventions into vernacular travels that operate from outside of the formulaic templates and hegemonic logi(sti)cs of the 'categorical imperative' imposed by the heuristics of the '(un)modern' and the '(non-)secular'. For this, we have to be attentive to how these travel practices engender new epistemic categories (for example, *ghumakkari, parikrama, jivanmukti,* etc.) and are rife with decolonial imaginaries that rethink the discursive limits of the prevailing conventions of theorizations, as well as nuance and deconstruct the historically contingent categories: 'India(n)', '(pre-)modern', and '(non-)secular'. Postcolonial scholarship in general falls short of problematizing the *universalizability* of these categories, but instead reappropriates them as axiomatic. In this connection, Vinay Lal (2012, p. 194) contends that postcolonialism remains 'unreflective about the imperialism of categories'. This brand of postcolonialism brings about a 'functional change in a sign system' (Spivak, 1988, p. 4), but the categories, theories, tools, and methods of knowledge production, Lal hints, are not 'engineered' for and with reference to the vernacular subjects of knowledge. Running the risk of some degree of generalization, I advance this provisional but provocative formulation to highlight the discursive limits of postcolonialism. In contrast, this volume uses 'decolonial thinking' (Lal, 2002, 2012; Nandy, 1983; Cohn, 1996; Mignolo, 1999a; 1999b, 2001, 2007, 2009) as a point of departure toward considering: what new categories do the vernacular travels invoke, and how do they equip us to make sense of the universe of alternative travel(ing) in question?

Toward this, the volume features eight essays that probe different aspects of vernacularity spanning the pre-colonial and the postcolonial lifeworlds. In chapter 2, Mattia Salvini Dharmavardhana Jñānagarbha works toward a theoretical framework for understanding travel from the point of view of Sanskrit Buddhism, while Lauren M. Bausch, in chapter 3, discusses 'ritual mobilities' – the exchange of ritual offerings between the heavenly and the earthly, from within the Vedic discourse. In chapter 4, Rafia Khan demonstrates how the patronage system in the medieval Indian courts from the thirteenth to fifteenth centuries impacted the attitudes of certain medieval 'traveling authors' toward sessility, mobility, and movement, which had implications not only for their careers but also for their political perspectives and literary production in general. In chapter 5, Martin Gaenszle looks into the lives of two important mendicant-ascetics in the late nineteenth and early twentieth centuries as case studies to illustrate how these 'un-modern' continued to travel as 'trans-border individuals' resisting, perhaps unwittingly, the statist processes of territorialization and sedentarization. In chapter 6, Swati Chattopadhyay juxtaposes two texts, Bankimchandra Chattopadhyay's novella, *Kapalkundala* (1866), and Rani Chanda's memoir of childhood days, *Amar Ma'r Baper Bari* (1972), to consider how the role of water travel and the description of coastlines, river passages and waterscapes vary across the texts. In so doing, Chattopadhyay explores the strategies and the vocabulary through which bibliomigrancy – movement between languages – and movement through space are memorialized in these works to trouble the (imperialist) idea of land as landscape. In chapter 7, Nilanjana Mukherjee and Gaurav Kumar show how literature from the Thar region, circulated through ephemeral oral performances, mobile musicians and bards, and several local versions of romances and epics, continue to mirror the nomadic-pastoralist lifeworld specific to that region. Chapters 8 and 9 discuss how the trope of homelessness acquires cultural significance in the postcolonial Indian context. In chapter 8, Ajay Skaria situates Indulal Yagnik's simultaneous love of home and homelessness in relation to the logic of transcendence and the politics of neighbourliness, within the coordinates of Indian nationalism. On the other hand, Vinay Lal, in chapter 9, argues that Gandhi's homelessness – Gandhi barely ever lived in a family home, housing himself instead in *ashrams*, and in 1936 he established himself in a desperately poor village in central India – can be read as a metaphor of his political resistance to the prevailing ethos of nationalism. Together, these chapters offer us glimpses of specific practices of vernacular travel that warrant fresh thinking.

This volume approaches these travels as expressions of indigeneity, though not untrammeled and untouched by the Western/outside epistemes. The agenda here is not to 'recover' indigenous modes of travelling as a gesture to 'write back to the empire', but to better understand the alternative conceptual universe from which they originate. This volume unpacks what makes these travels defy the log(ist)ics of the 'imperial categories' within the larger framework of vernacularity vis-à-vis the genealogies of structures of power and that of knowledge formations based on the universalization of categories, cultures and performativities. This approach, it is envisaged, will offer an 'epistemic break' in thinking through 'travel(ing) theory' by untangling it from the 'imperial categories'. This will then render possible inter- and intra-referencing among the vernacular performativities of travel, dispelling the (post)colonialist episteme as the dominant heuristic outfit for understanding such travels. It will thus foreground decolonial 'world-making', and, by extension, help decolonizing our historiographic, hermeneutical, and heuristic tools of inquiry.

Notes

1 This piece is a slightly reworked version of my article, 'Decolonizing Travel(ing) Theory: Vernacular Travels in the Indian Context', published in issue 124 of *Cultural Critique* (Summer 2024, pp. 164–176). A preliminary version of this essay was presented at the workshop on 'Migration, Journey, Travelling & Nostalgia: Travel Narratives in Hindi Literature' organized by the Department of South Asian, Tibetan & Buddhist Studies, University of Vienna. I thank Charu Gupta and Alessandra Consolaro for their feedback. Free from any linguistic connotation, I use 'vernacular travel' across this volume as a shorthand for any form of indigenous mobility – without definitional fixity or chronological finitude – that is, broadly speaking, the non-canonical travelling cults outside of the dominant discourse.

2 It is germane in this context to remember that 'millions of people still live outside 'history'. They *do* have theories of the past; they *do* believe that the past is important and

shapes the present and the future, but they also recognize, confront, and live with a past different from that constructed by historians and historical consciousness. They even have a different way of arriving at that past' (Nandy, 1995, p. 44; italics in the original).

3 Sankrityayan was a polyglot and polymath who travelled extensively and is credited with having established the travelogue as a genre in the Indian context. For this, he is fondly remembered as the 'father of Indian travelogue'. Yagnik was an independence activist, a writer, and a film-maker. He renounced his family and vowed to live like a *fakir* (mendicant). Sachchidananda Hirananda Vatsyayan, popularly known by his pen name Agyeya, was a literary practitioner, journalist, translator, and revolutionary. Gandhi possibly does not need any introduction. All four figures were ardent patriots and their practice of travelling, although I veritably term it 'vernacular', emerged from deep-seated anxieties concerning the nation-state, which in itself is a 'modern' category. This ambiguity/hybridity is characteristic of 'vernacular modernity' which I discuss in this essay.

4 Readers may profitably consult Mignolo's (1999a, 2009) concept of 'epistemic racism', or alternatively, Sousa Santos's (2014) concept of 'epistemicide', in this regard.

References

Ahuja, R. (2009). *Pathways of empire: Circulation, 'public works' and social space in Colonial Orissa, c. 1780–1914*. Orient Blackswan.

Alam, M., & Subramaniam, S. (2007). *Indo-Persian travels in the age of discoveries, 1400–1800*. Cambridge University Press.

Banerjee, S., & Basu, S. (2015). Secularizing the sacred, imagining the nation-space: The Himalaya in Bengali travelogues, 1856–1901. *Modern Asian Studies, 49*(3), 609–649.

Bhabha, H. K. (1994). *The location of culture*. Routledge.

Chatterjee, K. (1999). Discovering India: Travel, history and identity in late nineteenth and early twentieth century India. In D. Ali (Ed.), *Invoking the past: The uses of history in South Asia* (pp. 192–227). Oxford University Press.

Chatterjee, P. (1997). *Our modernity*. Sephis and Codesria Publication.

Chatterjee, P. (1998). Hundred years of fear and love. *Economic and Political Weekly, 33*(22), 1330–1336.

Cohn, B. (1996). *Colonialism and its forms of knowledge*. Princeton University Press.

Eck, D. L. (2012). *India: A sacred geography*. Harmony Books.

Fischer, L. (1962). *The life of Mahatma Gandhi*. Collier Books.

Foucault, M. (1972). *Archaeology of knowledge*. Tavistock.

Fricker, M. (2007). *Epistemic injustice: Power and the ethics of knowing*. Oxford University Press.

Gandhi, M. K. (1964). *The collected works of Mahatma Gandhi*. Publications Division, Ministry of Information and Broadcasting, Government of India.

Gupta, S. (2017). *Cultural constellations, place-making and ethnicity in Eastern India, c. 1850–1927*. Brill.

Gupta, C. (2020). Masculine vernacular histories of travel in Colonial India: The writings of Satyadev "Parivrajak". *South Asia: Journal of South Asian Studies, 43*(5), 836–859.

Hacking, I. (1999). *The social construction of what*. Harvard University Press.

Hegel, G. W. F. (2007). *The philosophy of history*. Cosimo.

Kerr, I. J. (2007). *Engines of change: The railroads that made India*. Praeger.

Lal, V. (2002). *Empire of knowledge: Culture and plurality in the global economy*. Pluto Press.

Lal, V. (2012). The politics of culture and knowledge after postcolonialism: Nine theses (and a prologue). *Continuum, 26*(2), 191–205.

Mayaram, S. (2003). *Against history, against state: Counterperspectives from the margins*. Columbia University Press.

Mbembe, A. (2001). *On the postcolony*. University of California Press.

Mignolo, W. (1999a). I am where I think: Epistemology and the colonial difference. *Journal of Latin American Cultural Studies: Travesia, 8*(2), 235–245.

Mignolo, W. (1999b). *Local histories/global designs: Coloniality, Subaltern knowledges and border thinking*. Duke University Press.

Mignolo, W. (2001). Coloniality of power and subalternity. In I. Rodríguez (Ed.), *The Latin American Subaltern studies reader* (pp. 224–244). Duke University Press.

Mignolo, W. (2007). Delinking: The rhetoric of modernity, the logic of coloniality and the grammar of de-coloniality. *Cultural Studies, 21*(2), 449–514.

Mignolo, W. (2009). Epistemic disobedience, independent thought and de-colonial freedom. *Theory, Culture and Society, 26*(7–8), 1–23.

Mukhopadhyay, B. (2002). Writing home, writing travel: The poetics and politics of dwelling in Bengali modernity. *Comparative Studies in Society and History, 44*(2), 293–318.

Nandy, A. (1983). *The intimate enemy: Loss and recovery of self under colonialism*. Oxford University Press.

Nandy, A. (1995). History's forgotten doubles. *History and Theory, 34*(2), 44–66.

Nigam, S. (1990a). Disciplining and policing the "criminals by birth", Part 1: The making of a colonial stereotype – The criminal tribes and castes of North India. *Indian Economic & Social History Review, 27*(2), 131–164.

Nigam, S. (1990b). Disciplining and policing the "criminals by birth", Part 2: The development of a disciplinary system, 1871–1900.*Indian Economic & Social History Review, 27*(3), 257–287.

Nijhawan, S. (2020). Journeying in the vernacular: Pilgrimage, tourism and nationalism in Hindi travelogues. In A. Ray & I. Banerjee-Dube (Eds.), *Nation, nationalism and the public sphere: Religious politics in India* (pp. 61–82). Sage.

Pratt, M. L. (1992). *Imperial eyes: Travel writing and transculturation*. Routledge.

Ray, A. (2019). Vicissitudes of reading the Mahabharata as history: Problems concerning historicism and textualism. *Journal of Literary Studies, 35*(2), 1–19.

Ray, A. (2020). The "Vagabond" as a nemesis of the tourist: Toward a postcolonial critique of Zygmunt Bauman. *Tourism Culture & Communication, 20*(2–3), 107–116.

Ray, A. (2021). *The Vagabond in the South Asian imagination: resilience, agency and representation*. Routledge.

Ray, A., & Banerjee-Dube, I. (2020). Nation, religion, identities – Crisscrossing concerns. In A. Ray & I. Banerjee-Dube (Eds.), *Nation, nationalism and the public sphere: Religious politics in India* (pp. 1–16). Sage.

Said, E. W. (1983). Traveling theory. In E. W. Said (Ed.), *The world, the text, and the critic*. Harvard University Press.

Sen, S. (2005). *Travels to Europe: Self and other in Bengali travel narratives 1870–1910*. Orient BlackSwan.

Sousa Santos, B. (2014). *Epistemologies of the South: Justice against epistemicide*. Paradigm Publishers.

Spivak, G. C. (1988). Subaltern studies: Deconstructing historiography. In R. Guha & G. C. Spivak (Eds.), *Selected subaltern studies* (pp. 3–34). Oxford University Press.

Walker, B. (1938). *Hindu world: An encyclopedic survey of hinduism* (Vol. 2). George Allen & Unwin.

The Pre-Colonial and the Scriptural

2 To go is to Know (That You Never Went)
A Sanskrit Buddhist Map of Selected Illusions

Mattia Salvini

anantamadhyam ākāśaṁ
buddhānāṁ caiva dharmatā |
tryadhvasamatikrānta
nirālamba namo 'stu te ||[1]

Without end or middle is space
And the nature of the Buddhas.
Gone beyond the three paths of time,
Supportless, homage to you!
      Mañjuśrī's praise of the Buddha,
      From the Ornament of the Light of Awareness Sūtra

Sanskrit verbal roots having the sense of movement or 'going to' (*gati*) can also have the sense of knowledge or awareness (*avagati/jñāna*): to 'go' is to 'know'.[2] This etymological mirror reflects Buddhist images of travels, paths, and vehicles, bringing them into view as illusions and impossibilities.

Travel figures in the transmission of knowledge and in its results; it is the horizon within which knowledge and education may be envisaged. Life is itself called a 'journey' (*yātrā*),[3] wherein travel allows for the accumulation of merit and awareness (*puṇya* and *jñāna*) that shall result in freedom from all suffering and from all ignorance. As teachers, the Buddhas, the Śrāvakas, and the Bodhisattvas travel to disseminate the Dharma.[4] As students, Bodhisattvas and other disciples travel to learn, to ask questions, and to pay homage to Buddhas and to the symbols of their presence. Buddhist stories of travel are primarily stories about seeking knowledge and merit, and sharing them. Travel is closely linked to education: this narrative fact echoes the semantic overlap between movement and understanding, 'going' and 'perceiving'. This feature of Sanskrit vocabulary manifests at opportune, instructive junctures.

Among the Sanskrit roots indicating movement, *gam* may be regarded as the most commonly used; I will initially focus on words derived from this root and then expand my analysis toward a conceptual 'decolonization' of Buddhist travel.

To begin with, though, I will address a premise of Buddhist thought.

DOI: 10.4324/9781003544524-4

Travel as a Metaphor of Nothing in Particular

Real travel occurs in space; metaphorical travel does not require space: this state-ment doesn't work too well in an illusory world.

Travel is clearly metaphorical when referring to activities that are inner and mental, not within the framework of physical presence and spatial location: 'going for refuge in the Buddha', 'going out of *saṁsāra*', etc., are not expressions refer-ring to actual movement. On the other hand, we may want to regard depictions of, say, the Buddha's travels to Śrāvastī, as *literal* travel involving movement through space. From a Buddhist perspective, however, such distinction may be less straightforward.

In a Buddhist world, it may be difficult to find a literal referent of 'travel'; where there is literally no movement, all travel is but a metaphor. This idea is, I believe, common to all Buddhist schools, although the details of its rationale may vary. There are no real persons – selves – that could be said to 'travel'; entities in general do not last long enough to move, since they are momentary.[5] Even the duality of movement and stillness is, for some Buddhists, no more than conceptual travail,[6] the fundamental trouble of literality, the ultimate *trepalium*.

Buddhas, as everyone else, move from one place to another only at the level of 'approximations of speech' (*upacāra*) that are in need of further interpretation (*neyārtha*). Such provisional expressions differ from those teachings where the meaning is explicit (*nītārtha*): selflessness conforms to the real characteristics of reality (*lākṣaṇika*). Narratives involving persons performing actions through time are indirections, as the listener/reader should eventually understand. When meta-phors are set aside, there are no real persons, no real movement and, according to some, no real time.[7]

The distinction between primary and metaphorical usages is itself an approx-imation or metaphor, a paradoxical compromise with ordinary perception, and this makes the term 'metaphor' helpfully elusive. Ordinary people misperceive travel from one location to another as 'real'; Buddhist teachings suggest oth-erwise. Their countless images of vanishing perception – of skies, mirages, flashes of lightning, foam – make it difficult to pinpoint the intent of words of traversable referents. Just as Buddhas engage with metaphorical travel and with metaphors of travel in order to guide others, so too ordinary beings may have to travel as much as it is necessary to understand that travel is as illusory as sentient beings and Buddhas.

This somewhat convoluted premise will hopefully become more comprehensi-ble, when applied to specific examples.

The Sugatas, the Supreme Travellers

Abstract terms, pressed by etymology, reveal their ancient roots still embed-ded in direct action. But the primitive metaphors do not spring from arbitrary subjective processes.[8]

Ernest Fenollosa

The term *sugata* is a ubiquitous epithet of the Buddhas. The prefix *su-* means something like 'well', while the participle *gata*, 'gone', is derived from the most common root indicating movement *-gam*. With some approximation, it has been rendered as 'Well Gone'.

The adjective *sugata* suffices, it is explained, to distinguish the Buddha from anyone else whomsoever; the Sugata has 'gone' (*gata*) in a uniquely praiseworthy (*praśastatva*), irreversible (*apunarāvṛtti*), and complete (*niḥśeṣa*) way, unlike anyone else. Therefore, only a Buddha is 'Well Gone' (*su-gata*).[9]

Non-Buddhist teachers travel on paths that stray into extreme views, while Buddhas proceed on the middle way. In this sense, only the Buddha, and not any other religious teacher, is genuinely praiseworthy (*praśasta*). This usage of *su-* is explained in analogy to the word *surūpa* ('someone with a good form'), where *su-* indicates praise of *-rūpa*.

Bodhisattvas, especially those on the higher grounds (*bhūmi*), resemble Buddhas in many of their qualities, but they have not gone beyond *saṃsāra* in such a way that they shall not return again to life and death. Thus, they have not gone in an irreversible way (*apunarāvṛtti*). When fever is completely destroyed, in such a way that it will not come back, it is called *su-naṣṭa*, an expression wherein *su-* indicates that something is irreversibly *-naṣṭa* (destroyed).

Arhats may have gone beyond *saṃsāra* irreversibly but they have not achieved the maximum level of perfection in their body, speech, and mind. Such perfection is a unique result of Buddhahood, and thus, their 'going' does not have completeness. They are not like a *completely* filled pot (*su-pūrṇa-ghaṭa*).

Only Buddhas can be called Sugata in all the three senses of the prefix *su*: they are the best among travellers.

The Tathāgatas, the Identical Travellers

All Buddhas travel in a near identical fashion. A Buddha's activities are not uniquely distinct; they are cosmic regularities repeated by all the Buddhas and even by Bodhisattvas on the path, with remarkable uniformity. The identical features of their manifestation are explicitly stated: all the Buddhas of this particular universe will be born in the same general area, and will accomplish Buddhahood in the same exact location, as part of a series of fixed deeds that are required so as to guide sentient beings.[10]

Most importantly, the realization of the nature of reality is the same for all the Buddhas: when it comes to omniscience, there can be no variation. This idea is embedded in the epithet *tathā-gata*, also derived from the root *gam*, and sometimes rendered as 'Thus Gone'. All the Buddhas have 'gone' 'thus' in a way that conforms to the nature of things, and that is identical for all the Buddhas; they have realized the undistorted nature of things.[11] Things vary, but they are identically illusory.

The Tathāgatas as Identical Freedom from Travel

The 'Tathāgata', Subhūti, is said to be someone that has not arrived from any-where, and has not gone anywhere; therefore, he is said to be the 'Tathāgata the Arhat, the Perfect, Complete Buddha'.[12]

The Sūtra wherein this passage occurs is primarily presenting the Buddha from the perspective of the 'Dharma-Body' (*dharmakāya*), which is ultimate emptiness free from arising or cessation, and the above statement should most likely be under-stood in that light: Mahāyāna texts often speak of the Buddha as identical with the nature of things that they have realized. The Buddha's 'going' is identical for all the Buddhas: the same applies to their 'non-going'. Thus, when Buddha Śākyamuni identifies himself with the previous Buddha Dīpaṁkara, this is explained as refer-ring to 'the sameness of the Dharmakāya'.[13]

What Arrives, and What Is Arrived at: The Dharma as *āgama* and *adhigama*

When using the term 'Dharma' as referring to one of the Three Jewels that consti-tute the refuge for Buddhist practitioners, it may be said in very general terms that it refers to the Buddha's teachings. A more precise characterization is sometimes offered, and it relies on two words derived from the root *gam*.

The *āgama-dharma* is the Dharma 'that has arrived' (*ā + gam*), i.e., the textual tradition, including aurally transmitted texts, disseminated by travelling *bhikṣus* or *dharmabhāṇakas* ('speakers of the Dharma'); the idea that a specific text 'goes forth in Jambudvīpa' (*jambūdvīpe pra + car*) recurs in Mahāyāna Sūtras describing their own destiny.[14] The *adhigama-dharma* is the Dharma 'that is arrived at', i.e., the realization of the paths and results that are taught within the textual tradition;[15] as we shall see, the process that culminates in *adhigama* often involves a signifi-cant amount of travel.

What Exists: The Paths of Time

One of the many possible Sanskrit words indicating a 'path' is *adhvan*. Within the lexicographical literature, this term appears within lists of synonyms for 'path', 'road', etc.[16]

Bhānujīdīkṣita explains the etymology of *adhvan* as follows:

atti balam | ada bhakṣaṇe ||
It eats (*atti*) one's strength. The root *ad* is used in the sense of 'eating'.

He furthermore quotes a half-verse from a different lexicon, wherein we find an explicit mention of an important Buddhist usage:

adhvā vā[17] *pathi saṃsthāne*
sāsravaskandhakālayoḥ ||[18]

adhvan is used for a path, a place,
for the aggregates with fluxes, and for time.

In Buddhist texts, the term is very often used to refer to 'the paths of time', meaning the past, the present, and the future, sometimes more specifically in the sense of past lifetimes, the present lifetime, and future lifetimes.[19] There is however a broader usage also expressed in the half-verse quoted above, wherein all the mental and bodily constituents of a non-liberated person (the 'aggregates with fluxes') can be called *adhvan*.

An even more encompassing usage is when all impermanent entities produced by causes and conditions (*saṃskṛta*), are called *adhvan*. This is explained by Vasubandhu in the following way:

> Those very conditioned things are *adhvan*s, because of 'being gone', 'going' and 'going in future', or in the sense that 'they are eaten (*adyante*) by impermanence'.[20]

Yaśomitra expands:

> Those themselves are *adhvan*s, etc.: the whole of Abhidharma is the meaning of the Sūtras, is the Sūtra-touchstone, is the explanation of the Sūtras; thus, even the *adhvan*s that are spoken of in the Sūtras, as synonyms of the aggregates, are here mentioned in the sense of 'gone', 'going' and 'going in future'. This *adhvan* is explained relying upon the *adhvan* as established in the world. Thus, in the world it is said: 'this *adhvan* has gone to the village, this *adhvan* goes, this *adhvan* will go'. In the same way, here to the '*adhvan* that has gone' is the past; the one that 'goes' is the present; the one that 'will go' is the future. 'They are eaten by impermanence'; 'the *adhvan*s' is the topic: this meaning is shown by relying upon the method of *nirukti*. 'They are eaten (*adyante*) by impermanence', they are consumed, thus they are '*adhvan*s'; thus, only conditioned things have been taught by the Bhagavat by using the word *adhvan*.[21]

The 'method of *nirukti*' is a way of deriving words on the sole basis of sound similarity without resorting to established derivational rules as found in Pāṇini's grammar. This approach is used rather often: it allows embedding desirable meanings in key terms, making them mnemonic rubrics that become immensely helpful when navigating a specific system of thought.

The above passage emphasizes the undesirable aspects of movement, paths, and travel: conditioned entities produced by assemblages of causes and conditions, have the nature of suffering precisely because they 'travel' through time, i.e., because of their impermanence. In turn, 'traveling' through conditioned entities means roaming through *saṃsāra*, from one state beyond one's control to another – this is often compared to being tossed in the ocean.[22]

Two concepts make 'travel' a suitable metaphor even in the context of ontology and cosmology: time and space – travel implies changing spatial location through time. A certain type of Buddhist 'travel', in turn, is meant to overcome the unavoidable discomfort of a mind involved with temporal or spatial location and refers to the entirety of Buddhist practice.

The Path beyond Time

The verse quoted at the beginning of this article praises the Buddha as 'having gone beyond the three paths of time' (*tri-adhva-samatikrānta*).

There are different possible interpretations. At a basic level, the verse could mean that the Buddha has transcended *saṁsāra*: 'dependent arising in twelve limbs' offers a summary of saṁsāric bondage that is commonly parsed into three 'paths of time' – past life, present life, future life. The three paths of time are also the main subject matter about which ordinary sentient beings develop 'views', and the teaching of dependent arising allows them to dissolve those views and become free from mental afflictions.[23]

Another possible interpretation that may resonate more closely with the Mahāyāna context of the verse is that it expresses the idea of 'non-placement' (*apratiṣṭhāna*) or 'no support'/'no point of reference' (*nirālamba*), thematized in the very Sūtra where the verse is found, as also elsewhere. This doctrine includes, and even emphasizes, non-placement in the three times. The following verse presents non-placement through a metaphor of travelling through the vastness of the ocean:

> Not on this shore, not on that shore,
> not placed in between them;
> the Prajñāpāramitā is understood
> in terms of the awareness
> of the sameness of the three paths of time.[24]

The translation as 'wisdom' may not convey that *prajñā* is related to *jñāna*, here translated as 'awareness'; in this and other contexts, such etymological connection is significant. The term *pāramitā*, commonly rendered as 'perfection', is often understood as related to 'the other shore' (*pāra*), and as literally meaning 'gone to the other shore' (*pāram* + *itā*). The above verse is playing on this meaning, explaining that perfect placement, or complete mastery, involves non-placement. Haribhadra explains the verse as follows:

> As for the Prajñāpāramitā that is regarded as being very close to the Buddhas and Bodhisattvas, due to their realization of the sameness of the dharmas belonging to the three paths of time, in terms of their aspect of non-arising; that Prajñāpāramitā, indeed, is not placed 'on this extreme', i.e., within *saṁsāra*, thanks to wisdom; not 'on that extreme', i.e., within *nirvāṇa*, thanks

to compassion; those two extremes have the characteristic, in due order, of eternalism and cutting-off; it is also not placed in the middle between those two – thus, it is not placed in *saṁsāra* or in *nirvāṇa*.[25]

The path that leads beyond the paths of *saṁsāra* (and also beyond the paths of *saṁsāra* and *nirvāṇa*) is referred to quite often by the term *pratipad* but even more often by *mārga*. Both terms appear in lists of the 'Four Truths of the Noble Ones', as the last Truth.[26] This is also identified as the Dharma – as being a comprehensive summary of Buddhist practice.

A Sample of the Buddhas' Travels

A Buddha is part of a cosmic regularity that, as discussed earlier, varies only within a certain range of fixed detail; certain activities – such as, importantly, awakening, turning the Wheel of Dharma and passing into Parinirvāṇa – are common to all the Buddhas and performed in rather similar ways.

Buddhas travel a lot; most of their travel is directly or indirectly related to the spreading of the Dharma. Śākyamuni Buddha's travels during his very last birth, from a certain perspective, are part of the series of fixed activities of all the Buddhas;[27] it is also worth pointing out that, according to certain Mahāyāna traditions, Śākyamuni Buddha had already attained Buddhahood by the time he had entered her mother's womb.[28] The following list of travels is tentative and not comprehensive:

> Descending from the Akaniṣṭha divine realm to Mahāmāyā's womb.[29]
> Visiting, with his friends, a farmer's village, where the Bodhisattva has a meditative experience.[30]
> Going to town, where the Bodhisattva encounters the four sights.[31]
> Escaping from the palace to the forest, on horseback.[32]
> Travelling to obtain teachings.[33]
> Travelling to practice in solitude.
> Walking to Vārāṇasī to turn the Wheel of Dharma.[34]
> Travelling to bless a place (for example, Khotan[35]).
> Sending innumerable *nirmāṇakāya*s.[36]

The above sample may suffice to demonstrate that a Buddha's travels are the visible manifestation of subtle processes related to acquiring and transmitting knowledge. All the travels of all the Buddhas are 'magical', in the sense that they are the necessary and partial surfacing of a layer of causality not accessible to ordinary and shared perception. The non-ostensible realm of karmic forces and mental dispositions is what determines the universe and its regularities and is perceptible by those with meditative insight. As for the ability to perceive causes and conditions in their entirety – the highest level of epistemic power and reliability – that is indeed unique to Buddhas:

> Even for a single moon-shape in a peacock,
> the causal assemblage in all its aspects
> cannot be known by those who are not omniscient:
> for, such knowledge is one of the strengths of the Omniscient Ones.[37]

More on the Magic of Travel: Divine Beings and Signs

The Buddha, right after awakening, remained in solitude in the very forest where he had meditated and attained Buddhahood; he expressed skepticism as to the existence of suitable disciples for the Dharma that he had realized. His first travel was prompted by two deities, who suggested that there are indeed some people who will be able to appreciate the Dharma for what it is – a *jewel*, which is not only precious and delightful but also inconceivably powerful.[38]

The theme of a divine (or semidivine) being prompting an important journey recurs in the hagiographies of numerous Buddhist masters: Śāntideva goes toward the South when urged by his own mother, an emanation of Vajrayoginī;[39] Maitrīpa sets on a journey to meet his Guru, Śavara, following indications that Tārā had offered him in a dream (Isaacson & Sferra, 2014, p. 427).

The last example may be called a *nimitta*, a sign, or an omen. Omens suggest hidden connections, and they are often occasioned by meditative practice – not unlike the appearance of divine and semidivine beings, past masters, Bodhisattvas, or Buddhas. In a more technical context, *nimitta* refers to a feature of the object of cognition, picked up by a mental function often translated as 'notion', which accompanies each and every moment of mind. According to one explanation, such features are called *nimitta* because they function as signs in that they allow the mind to infer that one has perceived a certain object rather than another; they stand at the juncture between cognition and the possibility of linguistic expression.[40]

Signs and the magical universe they presuppose figure prominently in the 'Letter on the Path that Goes to Potalaka' (Potalakagamanamārgapattrikā), a short text in the Tibetan Tengyur, attributed to the Bodhisattva Avalokiteśvara.[41] The pilgrim shall stop at appointed locations and chant specific mantras, observe the signs that arise, and then proceed accordingly. This will result in a visionary encounter with Avalokiteśvara, an encounter whose precise nature may vary (presumably due to karmic propensities and different levels of mental purification).

The prominence of magic reflects the fundamental Buddhist principle that mind has the primary causal role in the universe and furthermore, is not ostensible It places Buddhist travel, and the knowledge and education that one may obtain from it, in a world far removed from contemporary ideas of obvious and easily documentable exchanges of knowledge.

Travelling for Leisure? Not Quite

> My journey will be there, and strength, happiness, flawlessness, pleasant abiding.[42]
>
> > Buddhist prayer for 'knowing the right measure of food'.

Travel for the purpose of entertaining oneself (and others) is not absent from the Buddhist narrative traditions. This type of travel needs to be addressed, considering that wealthier contemporary societies often see travel primarily in that light. The very etymology of the term 'travel' (related to trouble, travail, and to the *trepalium* – an instrument of torture) should warn us against projecting this strong association between travel and diversion onto pre-modern cultures. Of course, economic migrants, refugees, etc. – a large portion of the human population – may still not automatically associate travel with leisure.

In the life of the Buddha, we have found two significant examples of travel for the sake of diversion: travelling to a farmer's village and travelling to town. In both examples, the final outcome differs from mundane diversion: the visit to the village occasions the Bodhisattva's experience of meditative absorption, while the sights of suffering that the Bodhisattva encounters in town elicit his wholehearted wish to find a solution to the suffering of *saṁsāra*. In both cases, travelling leads to crucial knowledge – to a perceptual shift of sorts.

Other such instances may be found in the Jātakas. In the story of Kṣāntivādin, a king travels to a forest to enjoy the scenery with the beauties of his harem, only to find himself implicated in a series of events that cause the Bodhisattva (Kṣāntivādin) to offer him a profound teaching on forbearance (*kṣānti*).[43] In King Harṣa's *Nāgānanda*, a leisurely stroll by the seaside becomes the occasion for the Bodhisattva's act of self-sacrifice.[44] What starts in the lighter vein of travelling for leisure takes a sudden dramatic turn, setting the stage for a Bodhisattva's act of magnanimity performed in pursuit of omniscience.

Conclusions: The Mind of Sanskrit Buddhist Travellers

> They shook, they stared as white's their shirt:
> Them it was their poison hurt.
>   'Terence, this is stupid stuff'/Alfred Edward Housman (1896).

There are different ways of looking at the project of 'decolonization', and I have elsewhere pointed out that I regard it primarily as a metaphor: a conceptual space for reconsidering what is (and what is not) at the centre of one's mental life (Salvini, 2019b).

This requires a certain level of inner freedom from prevalent institutional and intellectual notions of what 'must' be at the centre. To avoid taking uncritically for granted that an inherited framework is preferable, it may be necessary to consider one's level of immunity to the constraints and pressures of contemporary academia.

I would contend that Buddhism and Sanskrit Buddhist thought is not easily 'allowed' to be at the centre. Being a peripheral resource rather than a central force, it is hardly a viable starting point for other possible acts of decolonization. As a preliminary diagnosis, I will propose that a vaguely understood 'empiricism' (of a type antithetical to Buddhist thought) underscores much of what is written within the field of Buddhist Studies. The intellectual centre, that defines the fundamental epistemology, is no doubt 'Western'. This term may well be an empty signifier, but it is a powerful one, a self-fulfilling prophecy about method and identity (especially, identity as scholars).

I think of decolonization in nonpolitical terms due to a practical reason: I know far too little about colonial history to disentangle the complex net of agency that made it possible. It also seems to me that a strongly politicized approach misses the mark, forcing one within the same framework. There is even a risk of substituting a colonially determined intellectual centre with a nationally determined one. This substitution may not be heuristically helpful: the pre-colonial Buddhist world is a world of kingdoms (or similar geopolitical entities), not nations in the modern sense. Within the contemporary global layout of nation-states, the routes of pre-colonial Buddhist travel would be hard to imagine – whether on the ground or on a conceptual geography. Beneficial metaphors have no place in the divisive literality of contemporary geopolitical maps.

I focused on the inner aspects of decolonization and travel, and on levels of speech and reference: this is my attempt to articulate features of Buddhist thought within the rubric of 'travel'. I take the mental and metaphorical world to be an adequate starting point to overcome colonial limitations. Taking an inner rather than an outer perspective is here a new (yet very ancient) starting point for any further type of illusory travel, an alternative to the prevailing terms of the debate. A healthy immunity to poisons can be developed by careful and intentional exposure in bearable, measured doses. It may not require any 'real' travel, but it will require to be in control of dosage and timing.

Buddhist travel is an act of illusionism and transmission of knowledge. It is not constrained by contemporary notions of 'scientific' empiricism: the mind determines the world, rather than the other way around. Buddhist travel is 'magical', as it discloses deeper aspects of causality through signs and inner experiences elicited by its inherently worthwhile risks. Travel is also non-travel, ultimate knowledge, freedom from knowing:

> *na gatam nāgataṁ stutvā*
> *sugataṁ gativarjitam |*
> *tena puṇyena loko 'yaṁ*
> *vrajatāṁ saugatīṁ gatim ||*

> Neither gone, nor arrived –
> Having praised the Well-Gone,
> Who has no 'going',
> May by that merit this world
> Reach the 'going' of the Well-Gone.[45]

Acknowledgments

I thank Guru Bhikṣu Vāgindra Śīladhvaja, Mitradatta Supitṛ, Giuliano Giustarini, and Junyang Ng for very useful suggestions and corrections. Part of this work was funded by the Australian Government through the Australian Research Council

Discovery Projects funding scheme, within the project 'Ancient Today' (DP DP220100370); I thank the ARC and my team-mates – especially, Yasmin Haskell.

Notes

1 Jñānālokālaṁkārasūtra (The Study Group on Buddhist Sanskrit Literature, 2005).
2 See, for example, *gatyarthānāṃ jñānārthatvād api gataṃ jñātam ity api coktam* ∥ Dharmottarapradīpa (Malvania, 1971: 235; *gato jñātaḥ | gatyarthānām avagatyarthatvāt* ∥ Nyāyasudhā on Anuvyākhyāna 2.2.2.)
3 Commenting on the term *yātrā* appearing in the quote on knowing the write measure of food prefaced to this article, the Śrāvakabhūmi offers the following explanation: *yat tāvad bhuktvā jīvatīty evaṃ yātrā bhavati|* (Śrāvakabhūmi Study Group, 1998).
4 See, for example, the Mahāvastu, where the Buddha encourages monks to travel and teach, saying that he shall do the same: *caratha bhikṣavaś cārikāṃ mā ca duve ekena agamittha ∥ santi hi bhikṣavaḥ satvāḥ śuddhā alparajā alparajaskajātikā aśravaṇatvād dharmāṇāṃ parihāyanti | ahaṃ pi gaṃse yena uruvilvāyāṃ senāpatigrāmakaṃ jaṭilānām anukaṃpāya ∥* (Marciniak, 2019, p. 536).
5 On this, see Von Rospatt (1995).
6 On the refutation of the reality of movement and stillness, see especially the second chapter of the Mūlamadhyamakakārikā, 'The examination of what is gone to and what is not gone to' (*gatāgataparīkṣā*). With 'elaboration' I here render *prapañca*. The term has also been rendered as 'multiplicity' (MacDonald, 2015, Vol. II, pp. $2, 41, 42, 43, 46, etc.) and 'hypostathization' (Siderits & Katsura, 2013). For a discussion of this term see Salvini (2016b) and Salvini (2019a).
7 See the Mūlamadhyamakakārikā, chapter 19 (Siderits & Katsura, 2013).
8 (Fenollosa, 1936, p. 22).
9 This explanation (with some variations) is found in a number of different texts. See, for example, the Arthaviniścayasūtranibandhana (Samtani, 1971, p. 244, where only two of three meanings appear), the Pramāṇasamuccaya (Steinkellner, 2005, p. 1) and the Bodhicaryāvatārapañjikā (de La Vallée Poussin, 1901–1914, pp. 2–3).
10 For the regularity in the activities of Bodhisattvas see, for example, Abhidharmakośabhāṣya, chapter 4 (Pradhan, 1967).
11 See for example: *sa evam adhigatasarvajñajñāno bhagavān yathā dharmāṇāṃ tattvaṃ vyavasthitaṃ tathaiva aśeṣato gatatvād buddhatvāt tathāgata it yucyate* (Prasannapadā, chapter 22, de La Valle'e Poussin, L. (ed.), 1903–1913, pp. 431–432).
12 *tathāgata iti subhūte ucyate na kutaścid āgato na kvacid gataḥ tenocyate tathāgato 'rhan samyaksaṃbuddha iti |* (Vaidya, 1961, p. 88).
13 See: 'The intent to refer to sameness is as when he said: "I myself was at that time *Vipaśvin*, the Perfect Complete Buddha", referring to the property of non-distinction of the Dharmakāya'." *samatābhiprāyo yad āha | aham eva sa tasmin samaye vipaśvī samyaksaṃbuddho bhūvam ity aviśiṣṭadharmakāyatvāt |* Mahāyānasūtrālaṁkārabhāṣya 12.18 (Lévi, 1907, p. 83) (*samatābhiprāyo* is an emendation for *satatābhiprāyo* appearing in Lévi's edition).
14 See Skilling (2004).
15 See Abhidharmakośabhāṣyabhāṣya 8.39ab: *saddharmo dvividhaḥ śāstur āgamādhigamātmakaḥ | tatrāgamaḥ sūtravinayābhidharmā adhigamo bodhipakṣyā ity eṣa dvividhaḥ saddharmaḥ |* (Pradhan, 1967, p. 459).
16 See Amarakośa, Dādhimatha (1995, p. 114).
17 I emend *nā* of the printed edition to *vā*, as I cannot make sense of the verse with *nā*. The verse, in fact, appears between brackets, which makes me wonder whether this may be an addition, from memory, by Paṇḍit Śivadatta.
18 Amarakośa, Dādhimatha (1995, p. 115). In my understanding, this verse is not so plausibly interpreted by Paṇḍit Śivadatta, who does not take into account that *sāsrava* and

skandha are Buddhist terms, and explains them without any reference to the Buddhist concepts of five aggregates and 'having fluxes', i.e., being conducive to the insidious growth of mental afflictions (see note 1 on the same page). This mistake, it must be stressed, is perfectly understandable – yet instructive about the long-term loss of familiarity with Buddhist thought among Sanskritists at the time of the composition of the commentary (possibly around the beginning of the twentieth century, when even the Abhidharmakośabhāṣya would not have been available to the author).

19 See, for example, Mūlamadhyamakakārikā, chapter 19.

20 *ta eva saṃskṛtā gatagacchadgamiṣyadbhāvād adhvānaḥ adyante 'nitayatyeti vā* | Abhidharmakośabhāṣya 1.7 (Pradhan, 1967, p. 5).

21 *ta evādhveti vistaraḥ* | *sarvābhidharmaḥ sūtrārthaḥ sūtranikaṣaḥ sūtravyākhyānam iti sūtroktānām api adhvādīnāṃ skandhaparyāyarūpāṇāṃ iha vacanaṃ gatagacchadgamiṣyadbhāvād iti* | *lokaprasiddham adhvānam apekṣyāyam adhvā vyākhyātaḥ* | *tathā hi loke kathyate* | *ayam adhvā grāmaṃ gato 'yam adhvā gacchati ayam adhvā gamiṣyatīti* | *evam ihāpi gato 'dhvā yo 'tītaḥ gacchati yo vartamānaḥ gamiṣyati yo 'nāgata iti* | *adyante 'nityatayeti vādhvāna ity adhikṛtaṃ nairuktaṃ vidhānam apekṣyāyam artho darśitaḥ* | *adyante anityatayā bhakṣyanta iti adhvāna iti saṃskṛtā evādhvaśabdena bhagavatā deśitāḥ* | Yaśomitra's Vyākhyā on Abhidharmakośabhāṣya 21.21.

22 See the Jātakamālā 22.22: *anena puṇyena tu sarvadarśitām avāpya nirjitya ca doṣavidviṣaḥ*
|
    jarārujāmṛtyumahormi saṅkulāt samuddhareyaṃ bhavasāgarāj jagat || (Kern 1891, p. 50).

23 On this, see Salvini (2011).

24 *nāpare na pare tīre nāntarāle tayoḥ sthitā* | *adhvanāṃ samatājñānāt prajñāpāramitā matā* || Abhisamayālaṃkāra 3.1 (Stcherbatky and Obermiller, 1929, p. 15).

25 *traiyadhvikāṇāṃ dharmāṇām anutpādākāreṇa tulyatāvabodhād āsannībhūtā yā prajñāpāramitā buddhabodhisattvānāṃ matā prajñāpāramitā, sā 'parānte saṃsāre prajñayā kṛpayā parānte ca nirvāṇe yathākramaṃ śāśvatocchedalakṣaṇe tayor madhye'pi na sthiteti bhavaśamānavasthānam* | Abhisamayālaṃkārakārikāśāstravivṛti on kārikā 3.1 (Amano, 2000, p. 48) (I have modified the orthography to more standard conventions).

26 *tatra duḥkhanirodhagāminīpratipad āryasatyaṃ katamat? ayam eva samyagdṛṣṭyādir āryāṣṭāṅgo mārgaḥ* || Arthaviniścayasūtra (Samtani, 1971, p. 15).

27 See, for example, Ratnagotravibhāga 2.53–56 (Nakamura, 1961, p. 169).

28 See, for example, the Saddharmapuṇḍarīkasūtra, chapter 15 (Dutt, 1952, pp. 206–214).

29 Lalitavistara, chapter 6 (Vaidya 1958).

30 Lalitavistara, chapter 11 (Vaidya, 1958).

31 Lalitavistara, chapter 14 (Vaidya, 1958).

32 Lalitavistara, chapter 15 (Vaidya, 1958).

33 Lalitavistara, chapter $2 (Vaidya, 1958).

34 Lalitavistara, chapter 26 (Vaidya, 1958).

35 *'phags pa glang ru lung bstan pa zhes bya ba theg pa chen po'i mdo.* Toh 357, Degé Kangyur vol. 76 (mdo sde, ah), folios 220.b–232.a.

36 *buddhānām nirmāṇakāyo 'nyeṣu buddhakṣetreṣūtpadyate anyeṣūtpanno viharati anyeṣu parinirvāti na tu sarvalokadhātavas tena śūnyāḥ kadācid bhavanti* [...] Muktāvalī (Tripāṭhī and Negi 2001: 8; *na tu* is reported as the reading of manuscript GHA, while the printed edition has *na ca*, which seems to me less plausible; similarly, *kadācid* is reported as the reading of manuscript GHA, while the printed edition has *kathaṃcid*, which seems to me less plausible).

37 *sarvākāram kāraṇam ekasya mayūracandrakasyāpi* | *nāsarvajñair jñeyam sarvajña-balam hi tajjñānam* || Abhidharmakośavyākhyā on Abhidharmakośabhāṣya 37.37.

48 *uttiṣṭha vijitasaṃgrāma sārthavāhānigha vicara loke* | *deśaya sugata varadharmaṃ bhaviṣyaṃti dharmaratnasyājñātāraḥ*||Catuṣpariṣatsūtra (Waldschmidt, 1952, 1956,

1960). Note the expression *vicara loke*, 'go about in the world', and the metaphor of the Buddha as a caravan-leader carrying a precious gem (on this, see Salvini (2016a, pp. 247–250).
39 See the Śāntidevajīvanī, De Jong (1975, p. 168).
40 'And that specificity is called "sign"; for, it is by means of that that things can be inferred in terms of specificities'. *sa eva ca viśeṣo nimittam ity ucyate | tenārthānāṁ viśeṣato 'numīyamānatvāt |* Munimatālaṁkāra (Li and Kano, 2015, p. 109).
41 rGyud 'grel rG lxxii.51, see Chimpa and Chattopadhyaya (1990, p. 192, note 67).
42 *yātrā ca me bhaviṣyati balaṁ ca sukhaṁ cānavadyatā ca sparśavihāratā ca* |Śrāvakabhūmi (Śrāvakabhūmi Study Group 1998). In its Pāli version, this prayer is regularly chanted before meals by monastics in the Theravāda tradition, to this day.
43 See Kern (1881), pp. 181–43.
44 See the Fourth Act of Nāgānanda (Skilton, 2009).
45 Paramārthastava (Pandey, 1994, p. 116).

References

Amano, K. (Ed.). (2000). *Abhisamayālaṁkārakārikāśāstravivṛti*. Heirakuji Shoten.

Chimpa, L., & Chattopadhyaya, A. (trans). (1990). *Tāranātha's History of Buddhism in India*. Motilal Banarsidass. (Reprint: first edition Simla 1970).

Dādhimatha, Ś. (Ed.). (1995). *Nāmaliṅgānuśāsana Alias Amarakoṣa with the Commentary Vyākhyāsudhā or Rāmāśramī of Bhānuji Dīkṣita*. Chaukhambha.

de Jong, J. W. (1975). La Légende De Śāntideva. *Indo-Iranian Journal, 16*(3), 161–182.

de La Valle'e Poussin, L. (Ed.). (1901–1914). *Bodhicaryāvatārapañjikā: Prajñākaramati's Commentary on the Bodhicaryāvatāra of Çāntideva, Edited with Indices* (Vols. 983, 1031, 1090, 1126, 1139, 1305 and1399 of Bibliotheca Indica). Baptist Mission Press.

de La Valle'e Poussin, L. (Ed.). (1903–1913). *Mūlamadhyamakakārikās (Mādhyamikasūtras) de Nāgārjuna avec la Prasannapadā Commentaire de Candrakīrti*. Académie Impériale des Sciences.

Dutt, N. (Ed.). (1952). *Saddharmapuṇḍarīkasūtram. With M.D. Mironov's Readings from Central Asian Mss*. Asiatic Society.

Fenollosa, E. (1936). *The Chinese written character as a medium for poetry*. City Lights Books.

Housman, A. E. (1896). *A Shropshire Lad*. K. Paul, Trench, Treubner.

Isaacson, H., & Sferra, F. (2014). *The Sekanirdeśa of Maitreyanātha (Advayavajra) with the Sekanirdeśapañjikā of Rāmapāla: Critical Edition of the Sanskrit and Tibetan texts with English Translation and Reproductions of the MSS. With contributions by Klaus-Dieter Mathes and Marco Passavanti*. Università degli Studi di Napoli "L'Orientale"/Asien-Afrika Institut, Universtität Hamburg 2014 [appeared 2015]. Manuscripta Buddhica 2 = Serie Orientale Roma CVII.

Kern, J. (Ed.). (1891). *The Jātaka-mālā or Bodhisattvāvadāna-mālā of Āryaśūra*. Harvard University.

Lévi, S. (1907). *Mahāyānasūtrālaṁkāra, Exposé de la doctrine du Grand Véhicule* (Vol. I). Paris.

Li, X., & Kanō, K. (2015). Critical edition of the Sanskrit Text of the Munimatālaṁkāra, chapter 1 (fol. 48r5-58r5): Explanation of *skandha, dhātu,* and *āyatana* based on Candrakīrti's Pañcaskandhaka. *The Mikkyo Bunka, 234*.

MacDonald, A. (2015). *In clear words. The Prasannapadā Chapter One* (Vol. 2). Verlag der Österreischen Akademie der Wissenschaften.

Marciniak, K. (Ed.). (2019). *The Mahāvastu A New Edition. Vol. III*. The International Research Institute for Advanced Buddhology Soka University.

Malvania, D. (Ed.). (1971). *Dharmottarapradīpa: (being a Sub-commentary on Dharmottara's Nyāyabinduṭīkā, a Commentary on Dharmakīrti's Nyāyabindu).* Kashiprasad Jayaswal Research Institute.

Nakamura, Z. (1961). *Ratnagotravibhāga-mahāyānottaratantra-çāstra Compared with Sanskrit and Chinese, with Introduction and Notes.* Sankibo-Busshorin.

Pandey, J. S. (Ed.). (1994). *Bauddhastotrasamgrahaḥ.* Motilal Banarsidass.

Pradhan, P. (Ed.). (1967). *Abhidharmakośabhāṣya of Vasubandhu.* TSWS 8. K. P. Jayaswal Research Institute.

Salvini, M. (2011). The Nidānasamyukta and the Mūlamadhyamakakārikā: understanding the Middle Way through comparison and exegesis. *Thai International Journal of Buddhist Studies, 2,* 57–95.

Salvini, M. (2016a). Ratna: A Buddhist world of precious things. In F. M. Ferrari & T. W. P. Dähnhardt (Eds.), *Soulless Matter, Seats of Energy. Metals, Gems and Minerals in South Asian Traditions* (pp. 219–254). Equinox.

Salvini, M. (2016b). A clear and learned guide in reading Candrakīrti's Prasannapadā. *Bulletin de l'École française d'Extrême-Orient, 102,* 415–435.

Salvini, M. (2019a). Etymologies of what can(not) be said: Candrakīrti on conventions and elaborations. *Journal of Indian Philosophy, 47,* 661–595.

Salvini, M. (2019b). Decolonizing the Buddhist mind. *American Philosophical Association Newsletter, 19*(1), 22–25.

Samtani, N. H. (Ed.). (1971). *The Arthaviniścaya-sūtra and its commentary (Nibandhana).* K.P. Jayaswal Research Institute.

Siderits, M., & Katsura, S. (2013). *Nāgārjuna's middle way. Mūlamadhyamakakārikā.* Wisdom Publications.

Skilling, P. (2004). Jambudvīpe pracaramāṇaḥ: The circulation of Mahāyāna Sūtras in India. *Journal of the International College for Advanced Buddhist Studies, VII* (pp 73-87).

Skilton, A. (trans). (2009). *"How the nāgas were pleased" by Harṣa & "The shattered thighs" by Bhāsa.* New York University Press JJC Foundation.

Steinkellner, E. (Ed.). (2005). *Dignāga Pramāṇasamuccaya, Chapter 1. A hypothetical reconstruction of the Sanskrit text with the help of the two Tibetan translations on the basis of the hitherto known Sanskrit fragments and the linguistic materials gained from Jinendrabuddhi's Ṭīkā.* www.oeaw.ac.at/ias/Mat/dignaga_PS_1.pdf

Śrāvakabhūmi Study Group (The Institute for Comprehensive Studies of Buddhism, Taisho University). (Ed.). (1998). *Śrāvakabhūmi, The first chapter, revised Sanskrit text and Japanese translation.* Taisho University Sogo Bukkyo Kenyujo, 4.

Stcherbatky, T., & Obermiller, E. (Eds.). (1929). *Abhisamayālaṁkāra-prajñāpāramitā-upadeśa-śāstra.* Academy of Sciences of the USSR.

The Study Group on Buddhist Sanskrit Literature. (Eds.). (2005). *Jñānālokālaṃkāra.* Taisho University Press. See also the GRETIL version. 2004. Based on the edition by Takayasu Kimura, Nobuo Otsuka, Hideaki Kimura, and Hisao Takahashi, "Bonbun kotei 'Chikomyoshogon-kyo' - Sarvabuddhaviṣayāvatārajñānālokālaṃkāranāmamahāy ānasūtra." In *Kūkai no shiso to bunka [*A Felicitation Volume Presented to Prof. Kicho Onozuka on his Seventieth Birthday],* Tokyo: 1 (596) – 89 (508).http://gretil.sub.uni -goettingen.de/gretil/corpustei/transformations/html/sa_jJAnAlokAlaMkArasUtra.htm (last access: April 12, 2022).

Tripāṭhī, R., & Negi, T. (2001). *Hevajratantram: With Muktāvalī Pañjikā of Mahāpaṇḍitācārya Ratnākaraśānti.* Central Institute of Higher Tibetan Studies.

Vaidya, P. L. (Ed.). (1958). *Lalitavistara.* Buddhist Sanskrit Texts, Vol. 1. Mithila Institute.

Vaidya, P. L. (Ed.). (1961). *Mahāyānasūtra samgrahaḥ.* Mithila Insitute.

Von Rospatt, A. (1995). *The Buddhist Doctrine of Momentariness.* Franz Veiner Verleg.

Waldschmidt, E. (Ed.)., (1952, 1956, 1960). *Das Catuṣpariṣatsūtra, eine Kanonische Lehrschrift über die Begründung der Buddhistischen Gemeinde. Text in Sanskrit und Tibetisch, verglichen mit dem Pali nebst einer Übersetzung der chinesischen*

Entsprechung im Vinaya der Mūlasarvāstivādins. Auf Grund von Turfan-Handschriten herausgegeben und bearbeitet. Teil I–III. Berlin 1962 (Abhandlungen der Deutschen Akademie der Wissenschaften zu Berlin, Klasse für Sprachen, Literatur und Kunst, 1960/1).

3 Vedic Travel

The Agnihotra and Beyond

Lauren M. Bausch

The *agnihotrabrāhmaṇa*s in late Brāhmaṇa texts describe travel pertaining to Vedic ritual.[1] The Agnihotra is a reduced form of complex *śrauta* rituals. Every Āhitāgni is obligated to perform this ritual twice a day, every day, after kindling his fires. In fact, a *yajamāna* is only eligible to perform more elaborate *śrauta* rituals if he dutifully performs the Agnihotra, which consists of, among other activities: milking a cow, boiling the milk over the *gārhapatya* fire, ladling the milk, offering the milk into the *āhavanīya* fire, and reciting various formulae. While numerous modes of travel occur in the Agnihotra, the ideas relating to such movements travelled even further through Indian intellectual history; this affected how the Vedas were understood in the colonial period and in later Indology.

According to Sāyaṇa, the Brāhmaṇa texts contain both (1) *vidhi*, the injunctions for performing rites, and (2) *arthavāda*, explanatory statements about the origins of rites and mantras expressed by means of explanatory connections (*bándhu*) and myths. As Joanna Jurewicz (2016, p. 308) has observed, Vedic language employs concepts from daily life to refer to abstract ideas, which, as Deleuze and Guattari (1994, pp. 5, 28, 140) say of philosophy, are periodically enlivened to keep the concepts alive. Although both *vidhi* and *arthavāda* comprise the Brāhmaṇa texts, *arthavāda* imparts the following five kinds of ritual travel.

Travel in the Agnihotra

The first form of travel occurs at creation. The *agnihotrabrāhmaṇa*s narrate the origins of the Agnihotra ritual through various cosmologies. These cosmologies describe how, in the beginning, only the primal being existed until this one desired to be born or to be many. Usually the primordial being is called Prajāpati, who is often homologized with the mind, but sometimes it is expressed as *bráhman*, *púruṣa*, or *vā́c*.[2] The act of creation itself is a movement from the primal being, the original mind, and ground of potentiality to relative existence in a given *loka* which I define, following Jurewicz (2016, pp. 253–254, 373), as a sphere of experience – quite literally for sightseeing. A desire arises in the primal being, which bifurcates the one mind into something desired and one who desires. Thought follows desire, in which the primal being begins to conceive of self and other. Out of this mind, something is generated, but what is generated cannot survive as a separate entity

DOI: 10.4324/9781003544524-5

without some mechanism for making the *loka*s, spheres of experience, continuous. Otherwise, what was generated would immediately collapse back into the primordial being. This separation is expressed as a propping apart of the two world halves, i.e., heaven and earth, and in terms of a continuous ritual offering initiated by Prajāpati into himself, in which heaven and earth reciprocally exchange offerings.

In one cosmology, Prajāpati first generated Agni but was afraid because Agni, death, wanted to eat, and there was no food besides himself (*ātmán*).[3] The text explains that the earth existed in Prajāpati's mind, so when Agni turned toward him, Prajāpati offered two oblations into himself, from which the plants, sun, and wind arose. Prajāpati protected himself by creating this mechanism for the continuation of Agni. If Prajāpati is the original mind, then his firstborn, Agni, may be understood as the intelligence aspect of Prajāpati's mind that consumes what it conceives to be apart from itself. Eating secondarily expresses cognizing objects dualistically. In the cosmology, the plants, sun and wind stand for the three *loka*s, which are Agni's food. Plants are a metonym for the earth and the physical sustenance of man, whereas the sun is a metonym for the yonder world – the unmanifest locus of what man generates through perception and, as a result, his mental sustenance. The movement between these two *loka*s in terms of the offerings made by man sustains his relative life.

In a second cosmology, Prajāpati emitted *prajā́*, the creatures or anything generated mentally or physically. Then Agni decided to burn the other *prajā́*, who in turn tried to extinguish Agni.[4] Agni entered man with the agreement that man will maintain Agni in this world and Agni will maintain man in the yonder world. Agni and man are closely connected. In fact, in the Brāhmaṇa texts, Agni and the brahmin making the offering are said to be the same.[5] Agni Vaiśvānara resides in every man as his intelligence, the conscious element of man that sends and receives offerings between heaven and earth. What is consumed by Agni as the offering are the other *prajā́*.

In a third cosmology, the primal being is *bráhman*, who existed alone in the beginning until desire arose in the form of a thought to procreate itself.[6] Like Prajāpati, *bráhman* practices asceticism and, emitting from itself the *devatā́s*, it placed them in their respective worlds. Agni was sent to earth, the wind to intermediate space, and the sun to the sky. After creating the worlds, *bráhman* wanted to come down to experience relative existence, but first it had to figure out how to make the worlds continuous. He made the worlds continuous through name and form (*nāma rūpa*), which are called two *bráhman*s. 'Mind is the same as form (*rūpa*), for one knows form through the mind, [thinking] "This is form."'[7] Similarly, 'Speech (*vāc*) is the same as name (*nāma*), for by speech he utters a name'. According to this cosmology, the primal being considered relative existence to be something desirable and to be enjoyed. The mechanism employed to make the worlds continuous is here described in new terms. Instead of an oblation poured into the fire, now the exchange between heaven and earth consists of offering name and form. It is still an exchange of offerings between worlds, but more abstract and explicitly cognitive.

In short, the first aspect of travel in the Agnihotra is the movement of the primal being into relative existence, travelling from its true nature beyond desire and thought in order to experience itself as a creature in relation to other creatures. Agni stands for the intellectual function of the primal being in its journey through relativity. Agni is one and the same as both the primal being and as the sacrificer, who feeds on the other *prajá* to ensure his physical and mental continuity. As such, Agni is both death (JB 1.12, VādhS 3.32) and a procreator (KS 6.7: 56.15–20).

The second aspect of travel in the Agnihotra has to do with the movement of offerings between the worlds. By means of these offerings, the sacrificer makes his *loka*s and his two bodies. Brian Smith (1989, p. 82) describes human life as 'a process of ritually constructing and refining a self, an *ātman*'.[8] The Agnyādhāna, or kindling of the sacred fires, initiates a lifelong *ātmayajña* (ŚBK 1.2.2.10; cf. ŚBM 2.2.2.14–18). *Mādhyandina Śatapatha Brāhmaṇa* 11.2.6.13 defines the *ātmayājin* as the one who knows, '"This my (new) body [*aṅga*] is formed by that (body of Yajña, the sacrifice)"… made up of the Rik, the Yajus, the Sāman, and of offerings' (tr. Eggeling; *ātmayājī yo védedám me 'nenáṅgaṃ saṃskriyáta…sá ṛṅmáyo yajurmáyaḥ sāmamáya āhutimáyaḥ svargáṁl lokám abhisámbhavati*). This passage also speaks of the sacrificer freeing himself from his mortal body (*mártya- śárīra-*), from evil, like a snake sheds his skin: *sa yatháhis tvacó nirmucyétaívam asmān mártyāc chárīrāt pāpmáno nírmucyate*. Jurewicz (2019) gives a comprehensive account of the physical body and the ritual, immortal self in the Agnyādhāna from *Jaiminīya Brāhmaṇa* 1.1-2. In this context, the ritual fires are homologized with the *yajamāna*'s *prāṇāḥ*, lifebreaths said to be *devāḥ*, divine powers, i.e., the sense faculties.[9] *Jaiminīya Brāhmaṇa* 1.2 specifies that the Agnihotrin places the oblation 'in his own body, in these immortal life-breaths [*tad ātman nidhatta eṣv amṛteṣu prāṇeṣu*]' and 'makes for himself an immortal body here [*amṛtaśarīram idaṃ kurute*]' (tr. Bodewitz, 1973, p. 20).[10] According to JB 1.17, offering in the divine womb, i.e., the *āhavanīya* fire, he emits his *ātman*, which arises in yonder sun.[11] In *Kāṇva Śatapatha Brāhmaṇa* 3.1.9.3,[12] through offering the Agnihotra, the *yajamāna* forms (*sáṃskaroti*) his own *ātmán*, which the sun tells him is his body (*ātmán*) 'made of libations and made of merit' (*tám eṣa ādāyódayate sá pareṇāsyaitám ātmánaṃ sáṃskaroti sa yad āmúṃ lokam ety áthainam eṣá ādāyódayate tám eṣá āhutimáyaḥ sukṛtamáya ātmáhvayaty éhy ayáṃ ta ātméti*).[13] According to *Mādhyandina Śatapatha Brāhmaṇa* 11.2.2.5-6, the offering becomes his body in the yonder world, by which he is delivered from death.

What is offered from this world becomes the sacrificer's body in the yonder world and that body in turn sends offerings back that are received by the sacred fire cum sense faculties. The two bodies of the *yajamāna*, his material body (*ātmán*) and his undying body (ŚBK *ātmán*; ŚBM *aṅga*; JB *amṛtaśarīra*), correspond to *loka*s or 'spheres of experience'. According to Gonda (1966, p. 34), *loka* sometimes means a world or situation and other times lacks a definite location. *Lokāḥ* in the plural may 'denote the various states of existence in this world', which can overlap (60).[14] Through the ritual offering, the sacrificer constructs his 'world', i.e., the situation in which he finds himself. He endeavoured to make a *sukṛtáṃ loká* (ṚV 10.16.4d; Bodewitz, 2019, p. 107) obtained by doing *puṇyaṃ karma* (TB 3.3.10; Ibid., p. 384).

dhātúś ca yónau sukṛtásya loká íty āha | agnír vai dhātā́ | púṇyaṃ kárma sukṛtásya lokáḥ | agnír evaínāṃ dhātā́ | púṇye kármaṇi sukṛtásya loké dadhāti | syonáṃ me sahá pátyā karomíty āha |

(TB 3.3.10)

He says, 'And the world of merit is in the womb of the bearer'. He says, 'Agni verily is the bearer. The world of merit is meritorious karma. Agni alone is the one who bears those [merits]. He puts in the meritorious karma, the world of merit. I make a pleasant situation for me along with my wife'.

The successful sacrificer becomes *púṇyaloka* 'whose world … is *puṇya* or obtained by *puṇya*' (Bodewitz, 2019, p. 385).[15]

After Prajāpati first offered the Agnihotra, day and night emerged as separate, alternating entities. Just as Agni and *prajā́* are death, day and night are referred to as two repeated deaths (*punarmṛtyu*; JB 1.6, 1.13).[16] In the Brāhmaṇa texts, day and night are called 'the ocean that takes all' (KB 2.9: *samudro ha vā eṣa sarvaṃharo yad ahorātre*) and two impassable oceans that 'come and withdraw in a revolving, alternating way' (Bodewitz, 2003, p. 45).[17] Day and night may also be understood as *loka*s (ŚB 10.2.6.7; Gonda, 1966, p. 34) and their constant revolution is depicted as a wheel that maintains the continuity of the *loka*s (AiB 5.29; Keith, 1920, p. 254).[18] So long as a sacrificer does not directly see the rotation of day and night, his merit is exhausted.

ahorātre vaí parivártamāne púruṣasya sukṛtā́ṃ kṣiṇutaḥ sa yáthā rath- ena dhāváyan rátha cakré parivártamāne pratyavékṣetaiváṃ hāhorātré parivártamāne pratyavékṣate tásya há nāhorātre sukṛtā́ṃ kṣiṇuto 'kṣī́yaṃ ha yayati yá evám etad véda |

(ŚBK 3.1.9.3; cf. ŚBK 3.2.6.3)

Day and night rotating exhaust (√*kṣi*) the merit (*sukṛtá*) of a man. Just as driving in a chariot, he looks down at the two wheels rotating, so he looks down at day and night rotating. Then day and night do not exhaust his merit (*sukṛtá*). He who knows it in this way, wins what is inexhaustible (*akṣī́yá*).

In order to look down (*prati+ ava+ √īkṣ*), the sacrificer had to go up to the *svarga loka* twice a day by means of the Agnihotra.

What the sacrificer finds in heaven is his merit from past offerings, which have been stored in the sun. Sometimes, the Brāhmaṇa texts refer to this as *prajā́*, the sacrificer's offspring or children. His own creatures, generated through previous offerings, comprise the food that nourishes and sustains his conditioned life.

eṣa vāvá mṛtyur yá eṣa tápati tásmād yā áto 'rvā́cyaḥ prajās tā mártyā átha yāḥ párācyas tā́ amṛtās tásyaitásya mṛtyór imā́ḥ prajā́ḥ prāṇéṣu raśmíbhir abhíhitā yathā́śvo raśanayābhihitaḥ syād…sa vā́ éṣo 'staṃ yánn agním eva práviśati |

(ŚBK 3.1.9.1)

This one that heats is indeed death. Therefore, those *prajā* being on this side are going to die. And those on the far side are undying. These *prajā* are harnessed to the breaths by rays of light/reins, just as a horse would be harnessed by a rein…This one verily, setting, enters Agni alone.

Bodewitz (2003, p. 157 n. 2) interprets the corresponding passage in the Mādhyandina recension as follows:

The creatures are harnessed by the sun. The reins or the bridle are attached to the *prāṇāḥ*, which on the one hand denote the vital breaths, but on the other hand also the openings of the senses (mouth, etc.). These reins are the rays of light.

Every time the *prajā́* are harnessed to the breaths, the sacrificer, in the form of his *prajā́*, experiences death again in the yonder world. Since these *prajā́* constitute the sacrificer's body, they are not different from him. They constitute part of a metaphorical domain of birth in the *agnihotrabrāhmaṇa*s and describe, in narrative form, a Vedic causal process.

The two Agnihotra offerings in the morning and evening create an ongoing cycle in which Sūrya, i.e., the milk libation, is offered into Agni in the evening and Agni into Sūrya in the morning (ŚBM 2.3.1.36; JB 1.9; Bodewitz, 2003,: p. 146).[19] Like *Jaiminīya Brāhmaṇa* 1.8,[20] the *Śatapatha Brāhmaṇa* explains that the offering passes the night in the condition of an embryo before being born through the morning offering:

sū́ryo ha vā́ agnihotraṃ … sa vā́ éṣo 'staṃ yánn agním eva yóniṃ gárbho bhūtvā právisati taṃ gárbhaṃ bhávantam imā́ḥ sárvāḥ prajā́ ánu gárbho bhavantī́litā́ iva hi śératé 'saṃjānānā átha yad rátris tirá eva tát kurute tirá iva hi gárbhaḥ |

(ŚBK 1.3.1.1; cf. ŚBM 2.3.1.3-5)

Sūrya verily is the *agnihotra*.…Going to set, he [the sun] having become an embryo (*garbha*), enters the very fire, the womb (*yoni*). Following the one who becomes an embryo, all these *prajā* become an embryo, for they lie down as if requested, being unaware. Then the night just conceals that, for the embryo is hidden as it were.[21]

The setting sun enters the fire as an embryo enters the womb. In the same way, all the *prajā́* stored in the sun become an embryo, lying unaware (*asaṃjānānā*), hidden. A cause and effect transaction occurs between this and the yonder world, represented by Agni and Sūrya. The sun entering the fire in the evening represents part of the sacrificer's body from the yonder world, his *prajā́*, entering Agni, his physical body represented by the *gārhapatya* fire and breaths. Night conceals the embryo, which is incubating and unconscious as a kind of latent potential waiting to be born. This night, as we know from the rotation of day and night, represents part of the cycle of the sun, which, if unseen, enables the continuity of relative experience by exhausting the sacrificer's merit.

So, the second aspect of travel concerns the Agnihotrin's offerings, which constitute his bodies in this and the yonder world. The milk libation is homologized with the sun, a metonym for the yonder world and a metaphor for the sacrificer's second body, which takes birth in Agni, homologized with the sacrificer, during the Agnihotra. The offerings stored in the sun are sometimes called merit and sometimes called *prajā́*. The Agnihotrin's aim is to directly see the embryonic offering from the yonder world before the *prajā́* entering his senses are consumed in his sense faculties. If the *prajā́* take birth in his body, the light of *svar* is blocked and the *prajā́* are used up in dualistic perception by the faculties. Like a womb, the yonder world is said to be concealed by darkness:

saurī́bhyām r̥gbhyā́m támasā vā́ asaú lóko 'ntárhitaḥ sū́ryo vai támaso 'pahantā́ tád eténaiva támo 'pahátya svargáṃ lokám upasáṅkrāmati |

(ŚBK 5.4.1.7)

With verses dedicated to the sun [he chants]. The yonder world verily is concealed by darkness. The sun verily is the dispeller of darkness. Therefore, with just this, having dispelled the darkness, he steps over to the *svarga loka*.

After dispelling the darkness, the *yajamāna* goes to the yonder world.

The third aspect of travel in the Agnihotra is the journey to the *svarga loka* in each performance of the twice daily ritual (Smith, 1989, pp. 105–112). The Agnihotra is called the ship by which one crosses over day and night, constant *punarmr̥tyu*.[22] According to Smith (106), the sacrifice is like a chariot for the purpose of ascending to heaven: 'the Agnihotra should be performed after sunrise so that it will be like a chariot with both wheels' [AiB 5.30]. He says that the sacrifice is a sturdy vehicle by means of which one makes the trip to heaven.

In addition to gaining the perspective for seeing the rotation or exchange between this and yonder world, during the Agnihotra, the *yajamāna* sits between the *gārhapatya* and *āhavanīya* fires, which stand for this and yonder world, in the *Aitareya* and *Śatapatha Brāhmaṇa*s (AiB 7.12.7; ŚBK 3.1.11.1-6; cf. ŚBM 2.3.3.13-16). Sitting between the fires destroys his evil – evil passes over him – and he becomes one by whom evil has been destroyed (ŚBK 3.1.11.2: *tásya haivam ántareṇa yáto 'gnī́ pāpmā́nam ápahatas tám pāpmā́ nābhyátyeti só 'pahatapāpmā́ bhavati*). Between the fires cum *loka*s, he can see the exchange between them:

átha yád v evā́ntareṇopaviśáty eṣā́ vai naúḥ svárgyā yád agnihotraṃ tásyā etásyā nā́váḥ(nā́váḥ) svárgyāyāḥ kṣīráhotaivá nāvājas tām átaḥ prā́cīm ábhyajati tásyā áto 'dhiróhaṇam tām ató 'dhirohati |

(ŚBK3.1.11.3)

> Then that he sits in between is [because] this Agnihotra is verily the boat
> conducive to *svar*. Of that boat conducive to *svar*, the offerer of milk indeed
> is the helmsman. He drives toward to the east. Hence, the boarding of that
> [boat]. Hence he boards her [the boat].[23]

Sitting between the fires, the gateway to *svarga loka*, he reaches and is firmly
established in the *svarga loka* (3.1.11.5: *ha yád dhute púnar aíti tát svargé loke
prátitiṣṭhati*). Knowing in this way, his merit becomes inexhaustible and he
wins what is inexhaustible (3.1.11.6: *ha yád dhute púnar aíti tát svargé loke
prátitiṣṭhati...tásya hākṣīyáṁ sukṛtám bhavaty akṣīyáṁ ha jayati yá evám etad
véda*).

Directly seeing the sun, or reaching the world of heaven, was an integral part of
every Agnihotra performance (Bodewitz, 1973, p. 268). Various sacrificial actions
land the Agnihotrin in heaven.[24] *Aitareya Brāhmaṇa* 5.31 describes the sacrificer
being placed in heaven like an elephant taking him up with his trunk.[25] Similarly,
Jaiminīya Brāhmaṇa 1.11 depicts going to heaven like a man on an elephant seat
being lifted up (156). In the *Pañcaviṁśa Brāhmaṇa*, going to heaven and coming
back down to earth is as easy as climbing up and down a tree:

> Just as one, having climbed up to the top of a big tree, would get down by
> taking hold of branch after branch, so he gets down into this world (viz., the
> earth) by means of this (viṣṭuti), in order that he may get a firm support.
>
> (PB 3.6.2, tr. Caland, 1931, p. 36)

> In the manner as they climb from here (i.e. from the ground) onto a tree, in
> the same manner they descend from it: having ascended unto the world of
> heaven, they thereby regain firm support in this world.
>
> (PB 4.7.10, tr. Caland, 1931, p. 61)

The sacrificer is often said to see (PB 13.9.19: *svargaṁ lokam apaśyat*) or touch
(KB 24.8: √*spṛś*) *svarga loka* (Gonda, 1966, p. 91).[26] Knowingly being in contact
with or gaining the *loka*s transforms them into a safe situation without threat of
harm (ŚBM 3.3.4.3).[27] From a position of safety, man wanted to live a hundred
years on earth (TB 1.2.1.2),[28] while at the same time continuously knowing his true
nature as the impersonal primal being. The Agnihotrin lives very close to the source.

Travel to *svarga* is intimately connected with knowing and wholeness in the
Brāhmaṇa texts. Bodewitz (1973, p. 253) observed, 'The expression *ya evaṁ veda*
(or *evaṁ vidvān, evaṁvid*) is fairly current in the brāhmaṇas'. And Thite (1975,
pp. 236–237) observes, 'We always find in the Brāhmaṇa texts the growing importance of knowledge. The mere action cannot produce the expected or the promised result'. Knowing is connected with light, in particular, the light of the yonder
world. Not knowing was expressed as darkness. Knowing is also connected with
conquering the three worlds. Knowing the yonder world, the Agnihotrin becomes
sarvam, whole, like Prajāpati before creation (Gonda, 1955; Smith, 1989, p. 62).

So, the third aspect of travel pertains to ascending to *svarga*, i.e., accessing a sphere of experience conducive to *svar* throughout the day. The sacrificer can only do this when he does not indulge in the consumption of his *prajā* in his sense faculties. Through ritual practice, he must develop the ability to see the embryos of his past creatures as they are being born in the senses without letting them take over his awareness and cover *svar*. Going to *svar* entailed a certain kind of knowing that embraced the unmanifest, conscious of the dynamic exchange of offerings between heaven and earth that makes possible the continuity of relative existence.

The fourth kind of travel in the Agnihotra concerns the Agnihotrin going abroad. When he goes away from his home, he cannot offer into the physical sacred fires, so the Agnihotrin mentally performs the ritual abroad by offering into his breaths, which substitute for the fires. In both the *Śatapatha Brāhmaṇa* and *Jaiminīya Brāhmaṇa*, Yājñavalkya explains how to perform the Agnihotra while away from home. In the *Kāṇva Śatapatha Brāhmaṇa*:

sá hovāca námas te 'stu yājñavalkya vétthāgnihotrám sahástraṃ dadāmíti
tad ápy éṣo 'sti ślókaḥ

> *kíṃ svid vidvān právasty agnihotrī gṛhébhyaḥ*
> *kathám svid asya kāvyaṃ kathaṃ sáṃtato agníbhir íti*

> *yo jáviṣṭho bhúvaneṣu sá vidvān pravásan vide*
> *táthā tád asya kāvyaṃ táthā sáṃtato agníbhir íti*

tan mánasaivānte bhavati

> *yat sá dūráṃ paretyātha tátra pramādyati |*
> *kásmíntsāsya hutāhutir gṛhe yám asya júhvatīti*

> *yó jāgāra bhúvaneṣu víśvā rūpāṇi yó 'bibhaḥ*
> *tásmíntsāsya hutāhutir gṛhe yám asya júhvatīti*

tát prāṇā évaitām áhutiṃ juhvati tásmād vā āhuḥ prāṇā évāgnihotram íti |
ŚBK 3.1.4.4 (cf. ŚBM 11.3.1.4-8)

He [Janaka] assuredly said, 'May there be homage to you, Yājñavalkya! You know the Agnihotra. I give a thousand [cows]'. About that too, there is a verse:

'Is the Agnihotrin who goes away from his homes (= *gārhapatya* fire) knowing? How is his inspiration (*kāvya*)? How is he continuously connected with his fires?'

[Yājñavalkya said,] 'He who is the fastest among beings is found to be knowing while going away. Likewise is his inspiration. Likewise he is continuously connected with the fires'.

Ultimately, that is due to the mind alone.

'When having gone far away, then he becomes negligent (*pra+√mad*) there, in what is his offering (*āhuti*) offered (*hutā*) which they offer (*√hu*) for him at home?'

'He who has awakened (*√jāgṛ*) in beings, who bore all forms, in him is his offering, which they offer in his house'.

That is, in *prāṇa* alone they offer the offering. Therefore, they say that *prāṇa* itself is the *agnihotra*.[29]

The *Jaiminīya Brāhmaṇa*'s version of this teaching is very similar, but even more explicit about who is the fastest and who has awakened:

agnihotrī gṛhebhyaḥ | kathā tad asya kāvyaṃ kathā saṃtato 'gnibhiḥ || iti |
yad agnīn ādhāyāthāpapravasati katham asyānaproṣitaṃ bhavatīti |
sa hovāca vājasaneyaḥ
> *yo javiṣṭho bhuvaneṣu sa vidvān pravasan vide |*
> *tathā tad asya kāvyaṃ tathā saṃtato 'gnibhiḥ ||*
mana iti hovāca | mano vāva bhuvaneṣu javiṣṭham | manasaivāsyānapaproṣ-
itaṃ bhavatīti ha tad uvāca || atha hainam upajagau
> *yat sa dūraṃ paretyātha tatra pramādyati |*
> *kasmin sāsya hutāhutir gṛhe yām asya juhvati ||*
iti | sa hovāca vājasaneyaḥ
> *yo jāgāra bhuvaneṣu sa vidvān pravasan vide[30] |*
> *tasmin sāsya hutāhutir gṛhe yām asya juhvati ||*
iti | prāṇa iti hovāca | prāṇo vā bhuvaneṣu jāgaraḥ | prāṇa evāsya sā hutāhutir
bhavati | tasmād āhuḥ prāṇo 'gnihotram iti | yāvad dhy eva prāṇena prāṇiti
tasmād agnihotraṃ juhoti || JB 1.20

[Janaka:] 'Does the agnihotrin who stays away from his house have knowledge (of his agnihotra at home?)? How is there religious inspiration for him in that case? How does he remain connected with (his) fires?' (This means:) 'When having established his fires he thereupon leaves home, how is there no leaving home then?'

Vājasaneya said:

> He who is the quickest among the beings, is found to be the one that knows (i.e. keeps being aware of his agnihotra) while staying abroad. Thus there is religious inspiration for him then. Thus he remains connected with (his) fires.

Mind he meant by saying that. Mind certainly is the quickest among the beings. 'Mentally there is no leaving home for him' he thereby meant to say.

Then he recited to him:

> When he makes a journey and then neglects his duty (of performing a mental agnihotra) there, in what is that oblation offered by him which they keep offering for him at home?

Vājasaneya said:

> He who wakes in the beings is found to be the one that knows while staying abroad. In him that oblation is offered by him which they keep offering for him at home.
>
> Breath he meant by saying that. Breath wakes in the beings. In breath that oblation is offered by him. Therefore they say: 'the agnihotra is breath'. For as long as he breathes so long he offers the agnihotra.

(tr. Bodewitz, 1973, pp. 63–64)

In the Kāṇva recension, the travelling Agnihotrin goes away from his homes. Homes in the plural refers to the *gārhapatya* fire, which stands for the original house of the other *śrauta* fires, this world (earth), and the physical body of the Agnihotrin.[31] In the *Ṛgveda*, the term *javiṣṭha* often refers to the quickness of mind (ṚV 6.9.5: *máno jáviṣṭham*; 10.71.7: *manojavéṣu*, vs. 8: *mánaso javéṣu*),[32] which is exactly how the *Śatapatha* and *Jaiminīya Brāhmaṇa*s interpret it. *Mādhyandina Śatapatha Brāhmaṇa* 11.3.1.6 is close to the *Jaiminīya* reading: *mána évaitád āha mánasaísyā́napaproṣitam bhavatī́ti*. Eggeling translates, 'he thereby means the mind: it is owing to his mind that there is no staying away from home on his part'. According to Sāyaṇa, *yo javiṣṭhaḥ* refers to the *yajamāna*. So, this high-speed mind that knows and is open to the give and take of offerings between the worlds travels away from the ritual fires, one's *gārhapatya* fire, and physical body.[33] Bodewitz (1973, p. 65, n. 9) suggests that:

> Probably *vidvān* and *kāvyam* in the verse refer to the consciousness, the awareness of the sacrificer who every morning and evening has full knowledge of what is going on with his fires (i.e. his second self, his vital powers or breaths) at home.

Whether in physical proximity to the sacred fires or not, the Agnihotra offering is made in the one that has awakened in beings, i.e., *prāṇa*, which is homologized with Agni and the *yajamāna*.[34] Bodewitz (2003, p. 138) suggests that when the sacrificer goes abroad, the actual Agnihotra is performed at his home while his *prāṇa* temporarily substitute for the fires, forming a kind of mental Agnihotra. We saw in ŚBK 3.1.9.1 that 'These *prajā* are harnessed to the breaths by rays of light', which parallels the sun entering the fire in the ritual. With every inhalation the Agnihotrin receives an offering in the form of one of his creatures-turned-embryo from yonder world and with every exhalation he makes an offering back. Without the fastest mind, the travelling Agnihotrin becomes negligent of the offering when he is not carrying out the physical enactment of the ritual exchange. Elsewhere, the *Śatapatha Brāhmaṇa*, identifying the *yajamāna* with *prāṇa*, says that as long as the Agnihotrin breathes he performs the offering, but when his breaths are cut off, so is the Agnihotra.[35] The *Śatapatha* further explains that when the libation is offered in the breaths, they do not pour the existing libation into the fire.[36]

In short, the fourth aspect of travel in the Agnihotra involves the Agnihotrin going abroad, which is to say away from the sacred fires that he maintains throughout his life. If he has realized the fastest mind, then he remains aware of the offering made into *prāṇa* regardless of his proximity to the physical sacred fires.

The fifth mode of travel in the Agnihotra occurs when the Agnihotrin passes away.[37] Theoretically, by offering the Agnihotra, one is released from *punarmṛtyu*, but only if the sacrificer is *ya evaṃ veda*: 'This, then, is the release from death in the Agnihotra: and, verily, he who knows that release from death in the Agnihotra, is freed from death again and again' (tr. Eggeling) (ŚBM 2.3.3.9: *saiṣāgnihotré mṛtyor átimuktir áti ha vaí punarmṛtyúm mucyate ya evám etā́m agnihotré mṛtyor átimuktiṃ véda*). There is a grave consequence to knowing or not, as indicated in the *Śatapatha Brāhmaṇa*:

te yá evám etád viduḥ yé vaitat kárma kurváte mṛtvā púnaḥ sámbhavanti te sámbhavanta évāmṛtatvám abhisámbhavanty átha yá evaṃ ná vidur yé vaitat kárma ná kurváte mṛtvā púnaḥ sámbhavanti tá etásyaivā́nnam púnaḥ punar bhavanti |

(ŚBM 10.4.3.10)

And they who so know this, or they who do this [karma], come to life again when they have died, and, coming to life, they come to immortal life. But they who do not know this, or do not do this [karma], come to life again when they die, and they become the food of him (Death) time after time [*púnaḥ punar*].

(tr. Eggeling)

One who knows thus is released from repeated death, but not one who does not know thus. In the Brāhmaṇa texts, knowing involves the mind operating in a certain way, directly cognizant of the exchange between the worlds, in conjunction with knowing one's true nature.

The above cosmologies reveal that *púruṣa* generates and maintains Agni in this world, while Agni generates and maintains man in yonder world. Of course, both *loka*s not propped apart and both bodies combined are ultimately *púruṣa* in his original nature. After a man dies, they place his physical body on the funeral fire and the son becomes the father (ŚBM 2.3.3.3-6). The *Śatapatha* quotes ṚV 1.89.9, which says, 'A hundred autumns are now in front (of us), o gods, where you have made old age for our bodies, where sons become fathers. Do not harm our lifetime in the midst of our progress' (tr. Jamison & Brereton, 2014, p. 222). The Brāhmaṇa explains, 'for he…who is the son now becomes again the father. And therefore it is that he should establish the two fires' (tr. Bodewitz, 2003, p. 170).[38] The identity of the father and son pervasive in Vedic thought (JB 1.56; Bodewitz, 2019, p. 113), when considered in light of the philosophy of the Agnihotra, suggests that the father is the man performing *karma* on earth which generates *prajā́*, his offspring cum second body in the sun. When a knowing man dies, his breaths become undying, entering into his undying body that consists of his oblations, mind, breath, etc. (JB 1.2: *'mṛtatvaṃ gacchati ya evaṃ vidvān agnihotraṃ juhoti*).[39] But if he does not directly know this, his *prajā* generate a new man and, in this way, the sons become the father who then creates new *loka*s, bodies, and *prajā́*. In other words, *punarmṛtyu* continues.

Two passages in the *Jaiminīya Brāhmaṇa* present an early form of the two-path theory, which Jurewicz (2016, pp. 579–630) treats in great detail. In 1.18 (cf. 1.46–50), *prāṇa* ascends first and announces to the gods how much good and evil was done by the deceased: *sa he ya ttā* [sic] *devebhya ācaṣṭa iyad asya sādhu kṛtam iyat pāpam iti.* Then the seasons take him across to the sun, who asks, 'Who are you?' If the deceased gives his personal or family name, the sun gives the *ātman* that has been in him back to the deceased and the seasons grasp him by his feet and drag him away (Bodewitz, 1973, p. 54). Thereafter, day and night take possession of his world, which is to say he returns to conditioned, relative existence subject to *punarmṛtyu*. However, if the deceased answers the sun's question correctly, then he becomes the sun and stays in *svar*, returning to his true nature as Prajāpati, the original mind:

ko 'ham asmi suvas tvam | sa tvāṃ svargya svar agām iti | ko ha vai prajāpatiḥ | atha haivavid eva suvargaḥ | sa hi suvar gacchati ||

taṃ hāha yas tvam asi so 'ham asmi | yo 'ham asma sa tvam asya ehi iti | sa etam eva sukṛtarasam apyati |

(JB 1.18)

PM: Needs to be done per author's reply to our query in Chapter 2

'Ka (who) am I, thou art heaven. As such I have gone to thee, the heavenly heaven'. Prajāpati is indeed Ka and he who knows thus is *suvargas* (heaven; sun). For he goes to heaven (*suvar gacchati*).

To him he (the sun) says: 'Who thou art, that one am I. Who I am, that one thou art. Come'. He approaches that essence of good deeds [*sukṛtarasam*].

(tr. Bodewitz, 1973, pp. 54–55)

At this point, the exchange between the two worlds that makes the *loka*s continuous stops and the Agnihotrin is said to go to the sun (ŚBM 1.9.3.9-16), obtain the *ātman* in the sun (JB 1.18), get *salokatā* or coexistence in the same world as the sun (JB 1.50), or enter the sun and the essence of his merits (JUB 3.14.1-9). *Punarmṛtyu* is defeated and personal, relative existence ends.

In short, the fifth aspect of travel in the Agnihotra is the trip that the deceased Agnihotrin makes to the sun, his second body (*ātman*), where he is asked about his identity. If he remains attached to his personal identity and does not know who he truly is, then his past offerings – also called *prajá̄*, *sukṛtá*, and *ātmán* – become subject to another round of birth and death. If he knows that he is, in fact, Prajāpati, he enjoys omniscience as the primordial being. Knowing oneself determines one's destiny. At the height of ritual practice in the Brāhmaṇa texts, the Agnihotrin expected that his sojourn to earth to experience relative existence would last a full 100 years but not go beyond that. It was only when sacrificers began to fail to directly know the *yajñá* and their true nature that Vedic teachings had to be enlivened and elaborated to explain what happens when the sacrificer does not know and, as a result, experiences repeated death over and over again.

The Agnihotra's Journey in Indian Intellectual History

The ideas related to these five mobilities in the Brāhmaṇa texts travelled in Indian thought, beginning with the Upaniṣads. The Upaniṣads constitute an important part of the Brāhmaṇa texts, appearing at the end of the Brāhmaṇa text of their respective Vedic *śākhā*. In this way, the teachings of the Upaniṣads continue the thought of the Brāhmaṇa texts, but evince more abstract language and perhaps even different audiences. For example, the cosmologies of the *agnihotrabrāhmaṇa*s feature Prajāpati and *bráhman* as the primordial beings, but the Upaniṣads focus more on *bráhman* and *ātmán*. While Prajāpati is still the father of the *deva*s and *asura*s, *bráhman* creates Prajāpati in BĀUK 5.5.1. Just as the cosmology in the *Śatapatha* featuring *bráhman* as the primordial being is said to come down to the worlds by means of name and form, which are two *bráhman*s, so in the *Bṛhadāraṇayaka Upaniṣad*, *bráhman* is described as twofold: embodied (*mūrta*) and unembodied (*amūrta*):

> *dve vāva bráhmaṇo rūpé | mūrtáṃ caivā́mūrtaṃ ca | mártyaṃ cāmṛ́taṃ ca | sthitáṃ yác ca | sác ca tyác ca ||*

> (BĀUK 2.3.1)

There are two forms of brahman: the embodied and the unembodied, the mortal and the immortal, the steady and the moving, the existing and that [imperceptible] one.

The embodied is explained as going to die, like a sense faculty, whereas the unembodied is undying, the breath in the body (*ātmán*) (BĀUK 2.3.1-6). Instead of the mortal and undying bodies found in the *agnihotrabāhmaṇa*s, in the Upaniṣads, the embodied is contrasted with the unembodied. The word for body in BĀUK 2.3.4-5 is *ātmán*.

In the *agnihotrabrāhmaṇa*s, *ātmán* was twofold, such that the Agnihotrin had both a physical *ātman* or body consisting of the *prāṇāḥ*, the sense faculties, in this world as well as a body in the yonder world that likewise consists of sense faculties, offerings, and merit. According to the *Jaiminīya Brāhmaṇa*, the undying body is made of numerous things:

so 'ta āhutimayo manomayaḥ prāṇamayaḥ cakṣumayś śrotramayo vāṅmayo ṛṅmayo yajurmayas sāmamayo brahmamayo hiraṇmayo 'mṛtas sambhavati |

(JB 1.2)

He from this becomes immortal in the form of an oblation, mind, breath, sight, hearing, speech, *ṛc, yajus, sāman, brahman*, and gold.

(tr. Bodewitz, 1973, p. 20)[40]

The two bodies exchange offerings, enabling the continuity of relative existence within the *loka*s. Both the Brāhmaṇas and the Upaniṣads describe the *ātmán* as consisting of vital functions, though the *Bṛhadāraṇyaka Upaniṣad* adds mental states as well:

sa vā́ ayám ātmā bráhma vijñānamáyo mánomáyo prāṇamáyaś cakṣurmáyaḥ śrotramáyaḥ pṛthivīmáya āpomáyo vāyumáya ākāśamáyas tejomáyo 'tejomáyaḥ kāmamáyo 'kāmamáyaḥ krodhamáyo 'krodhamáyo dharmamáyo 'dharmamáyaḥ sarvamáyaḥ |

(BĀUK 4.4.5)

This *ātman* verily is *bráhman*, which is made of consciousness, made of the mind, and made of breath, made of the eyes, made of the ears, made of the earth, made of water, made of wind, made of space, made of fiery energy, made of the absence of fiery energy, made of desire, made of the absence of desire, made of anger, made of the absence of anger, made of dharma, made of the absence of dharma, made of everything.

Like the Brāhmaṇas, the Upaniṣads still consider the *ātmán* to be made of speech, mind, and breath (BĀUK 1.5.3: *etanmáyo vā́ ayám ātmā́ | vāṅmáyo mánomáyaḥ prāṇamáyaḥ*), but these discourses move toward consolidating these aspects of *ātmán* into a supreme self beyond the senses, from which all karma arises (BĀUK 1.6.3: *átha kármaṇām ātméty etád eṣām uktham | áto hi sárvāṇi kármāṇy uttíṣṭhanti*). In the *Bṛhadāraṇyaka Upaniṣad*, Yājñavalkya explains *bráhman*, identified in this passage with the *ātmán* in all beings, as follows:

ná dṛṣṭér draṣṭāram paśyer ná śrúteḥ śrotāram śṛṇuyāḥ ná mater mantāram manvīthā ná vijñātér vijñātāram víjānīyā eṣá ta ātmā́ sarvāntaró 'to 'nyad ārtam |

(BĀUK 3.4.2; cf. 4.3.23-31, 2.4.14)

You could not see the seer of sight. You could not hear the hearer of hearing. You could not think the thinker of thought, you could not know the knower of what is known. This is your *ātmán* that is in everything. Anything other than this is afflicted.

In the *Bṛhadāraṇyaka*, the *ātmán* is said to be in all beings, but is no longer explained as what enables the continuity of the worlds; instead, it is behind all sensory experience, yet beyond sensory experience.[41] According to the *néti néti* ('not' 'not') doctrine, *ātmán* is beyond thought and language such that anything one could say or think about it would not be true because it is not graspable by anything (BĀUK 2.3.6, 3.9.26, 4.2.4, 4.4.22, 4.5.15).

Like the Brāhmaṇas, the Upaniṣads still refer to *ātmán* as the physical body (BĀUK 1.2.3, 1.2.7, 2.3.4-5, ŚU 2.11, etc.) and describe the *ātmán* in terms of the worlds (BĀUK 1.2.7: *tásyemé loká ātmā́naḥ*; 1.4.16: *ayaṃ vā́ ātmā sárveṣāṃ bhūtā́nāṃ lokaḥ*). However, the Upaniṣads increasingly refer to *ātmán* as the undying self in the heart.[42] The heart increasingly replaces the yonder world as another metaphor for the locus of the unmanifest *ātmán*.[43] In addition, the self is said to be inside the body (KaṭhU 4.12, MuṇḍU 3.15).

Nevertheless, traces of the two *ātmán* doctrine occur throughout the Upaniṣads alongside various other kinds of *ātmán*s. In *Bṛhadāraṇyaka Upaniṣad* 4.3.35, 'Just as a fully loaded cart let loose would go, so this corporeal body (*śārīrá ātmā́*) ridden upon by the knowing body (*prājñéna ātmánā*) let loose goes' (*tad yathā́naḥ súsamā́hitam utsárjaṃ yāyā́d evám évāyáṃ śārīrá ātmā́ prājñénātmánānvārū́ḍha utsárjaṃ yāti*).[44] In *Chāndogya Upaniṣad* 8.7-12, Prajāpati teaches Virocana and Indra the *ātman* by first saying it is the body (*ātman*) in 8.8.1 and 4. Virocana accepts his teaching, but Indra continues to investigate the *ātman* until he knows that this physical body (*śarīra*) is the mortal seat of the immortal and bodiless (*aśarīra*) *ātman*. Section 12.1 contrasts the embodied (*saśarīra-*) with the unembodied (*aśarīra-*). This narrative uses *ātman* in the sense of both body and self. Similarly, in the *Aitareya Upaniṣad*, *ātman* occurs in the sense of the primordial being who created everything in chapter 1, what is corporeal in chapter 2, and the *prajña- ātman*, through which one becomes immortal, in chapter 3 (Cohen, 2018, p. 273). Cohen suggests that this progression is a structural device of the text. *Kauṣītaki Upaniṣad* employs *ātman* in the sense of body (2.12, 4.2, 4.10) and the inner essence or immortal self (Ibid., p. 278). Interestingly, *ātman* is used in the sense of body numerous times in the *Śikṣāvallī* of the *Taittirīya Upaniṣad* (TU 1.3.1, 1.3.4, 1.5.1, 1.7.1), but rarely in the sense of 'inner self'.[45] *Kaṭha Upaniṣad* 2.22 contrasts the great (*mahat*), bodiless (*aśarīra-*) *ātman* with the phenomenal body (*śarīra-*), but says in 3.11 that the unmanifest (*avyakta*) is higher than the *mahat ātman* and, moreover, *puruṣa* is higher than the unmanifest. *Kaṭha Upaniṣad* 3.13 distinguishes between an individual, cognitive self (*jñāna- ātman*) – who sees, touches, thinks, knows, and acts – and a universal cosmic self (*mahat ātman*); both of these are transcended in the tranquil self (*śānta- ātman*) (Hirst, 2018, pp. 116–118). In the *Muṇḍaka Upaniṣad*, *ātman* is not as prominent as *puruṣa* and *brahman*, but wise ascetics go 'through the door of the sun' to where the immortal *ātman* is (MuU 1.2.11; Cohen, 2018, p. 311). There are two *ātman*s in the *Śvetāśvatara Upaniṣad*, which says that the *ātman* is grasped in the body

(2.15: *ātmātmani gṛhyate*).[46] Signe Cohen (p. 344) observes, 'The *Maitrī Upaniṣad* differentiates between two different *ātmans*: The highest *ātman*, who does not act and is merely a spectator, and the lowest *ātman*, called *bhūtātman* ("the creature *ātman*"), which is that which is reborn and suffers karmic consequences'.[47] The *Maitrī Upaniṣad* (7.11) concludes: 'For the sake of the experience of truth and untruth the great *ātman* becomes two' (tr. Ibid., p. 344). These examples show that different concepts of the *ātman* circulated in the Upaniṣads and these concepts presented numerous variations on the philosophy of the Brāhmaṇa texts.

Later commentators like Śaṅkarācārya and Sāyaṇa use the terms *jīvātman* and *paramātman* to describe the two *ātmán*s. While these terms are not found in the earliest Upaniṣads, phrases like *parama-brahman* (BĀUK 4.1.2-7) and *parama-loka* (4.3.20) occur. Instead, the earliest Upaniṣads express these ideas using other terms. For example, *Chāndogya Upaniṣad* 8.3.4 states:

atha ya eṣa samprasādo 'smāc charīrāt samutthāya paraṃ jyotir upasam-padya svena rūpeṇābhiniṣpadyata eṣa ātmeti hovācaitad amṛtam abhayam etad brahmeti |

(ChU 8.3.4)

The perfectly serene one, rising up from this body (*śarīra*), reaching the highest light, remains established in his own nature. This is the *ātman*. This is immortal. This is beyond all fear. This is *brahman*.

In his commentary, Śaṅkarācārya glosses rising up from the body with the idea of giving up the identity of the self with the body. He understands the highest to mean the *paramātman* and light to mean the nature of consciousness (*paraṃ paramātmalakṣaṇam vijñaptisvabhāvaṃ jyotir*). Similarly, explaining *Chāndogya* 6.16.3, he equates understanding the self in terms of the body with identifying the self with the *jīvātman*.[48] The allegory of the two birds in one tree from *Ṛgveda* 1.164.20-22 occurs in *Muṇḍaka Upaniṣad* 3.1.1 and *Svetāsvatara Upaniṣad* 4.6-7. In his commentary on the *Ṛgveda*, Sāyaṇa interprets the two birds to be the *jīvātman* and the *paramātman* in the body (*deha*).[49] The *jīvātman* enjoys the fruit or rewards of actions, while the *paramātman* is a passive spectator. Śaṅkarācārya also glosses tree as body in *Muṇḍaka Upaniṣad* 3.1.1. It is interesting that here *jīvātman* is considered to be *in* the body rather than the body itself. According to this interpretation, the two *ātman*s dwell in one body.

In the *agnihotrabrāhmaṇa*s, Agni is homologized with breath and the *prāṇāḥ* (breaths) are called *deva*s, but imply the sense faculties. In the *Bṛhadāraṇyaka Upaniṣad*, breath is seen as the best of the sense organs (*karman*) because death cannot exhaust it and all the senses are forms of it (1.5.21; cf. 6.1.7-14).[50] The seven *ṛṣi*s are the breaths, consisting of the two ears, two eyes, the two nostrils, and speech (2.2.3-4; Jurewicz, 2016, pp. 332–348). And yet, the *ātman* remains inside breath, taking breath as a body (*śarīra*); the same goes for the other sense faculties (BĀUK 3.7.16-23).[51] Yājñavalkya identifies *prāṇa* with *brahman* (3.9.9, 4.1.3, 4.4.7). *Prāṇa* gives rise to all this, and all beings are yoked to vital breath

(5.13.1-2). So Agni, prominent and homologized with *prāṇa* in the Brāhmaṇa texts, loses importance and is replaced with *prāṇa* in the Upaniṣads.

Kāṇva Śatapatha Brāhmaṇa 3.2.5.3 understood, 'Mind is the same as form (*rūpa*), for one knows form through the mind, [thinking,] "This is form."' The same is true in the Upaniṣads (BĀUK 1.5.3). But according to *Bṛhadāraṇyaka Upaniṣad* 3.9.20, the sun is established in the eye, for one sees by means of the eye, and forms are established in the heart, for one knows forms by means of the heart. Like this, the *manas* of the *Śatapatha* is sometimes replaced with *cakṣu* and *hṛdaya* in the Upaniṣad:

> *kíṃ deváto 'syāṃ prā́cyāṃ díśy asī́ty ādityádevata íti | sá ādityaḥ kásmin prátiṣṭhita íti | cákṣuṣī́ti | kásmin nu cákṣuḥ prátiṣṭhitam íti rūpeṣv íti cákṣuṣā hí rūpā́ṇi páśyati | kásmin nú rūpā́ṇi prátiṣṭhitānī́ti hṛ́daya íti hovāca | hṛ́dayena hí rūpā́ṇi jānā́ti | hṛdayé hy evá rūpā́ṇi prátiṣṭhitāni bhavantī́ty evám évaitád yājñavalkya ||*

(BĀUK 3.9.20)

> 'What is your deva when you are in this eastern direction?' '[I am one who has] the sun (*āditya*) as a deva'. 'Where is the sun established (*pratiṣṭhita*)?' 'In the eye'. 'In what is the eye established?' 'In forms, for one sees forms by means of the eye'. 'In what indeed are forms established?' 'In the heart', he said, 'For, one knows forms by means of the heart. For indeed forms have been established in the heart'. 'It is indeed so, Yājñavalkya'.

Here materiality is experienced, thanks to the sense organ and sense object, which are ultimately established in the heart, whereas in the Brāhmaṇa, mind encompassed both functions.

The Upaniṣads add a new element to the faculties animated by the *ātmán* that is not found in the Brāhmaṇa texts, namely consciousness (*vijñā́na*). The *ātmán* in the *Bṛhadāraṇyaka Upaniṣad* is described as: 'That *púruṣa* which consists of consciousness, the inner light in the breaths, in the heart'.[52] In the analogy of rock salt dissolved in water at *Bṛhadāraṇyaka Upaniṣad* 2.4.12, the great being is boundless, a mass of consciousness (*vijñānaghaná*). After going through the various faculties in the body, including breath and speech, the eyes and the ears, mind and skin, the *Bṛhadāraṇyaka Upaniṣad* describes the *ātmán* in terms of *vijñā́na*:

> *yó vijñā́ne tíṣṭhan vijñā́nād ántaro yáṃ vijñā́naṃ na véda yásya vijñā́naṃ śárīraṃ yó vijñā́nam ántaro yamáyaty eṣá ta ā́tmāntaryā́my amṛ́taḥ |*

(BĀUK 3.7.22)

> He who remains in consciousness, inside of consciousness, whom consciousness does not know, who has consciousness for a body, who being inside of consciousness controls, he is your *ātman*, the undying inner controller.

In the Brāhmaṇa texts, Agni serves the intellectual function of one's intelligence or consciousness. In the Upaniṣads, an abstract concept is designated to serve this function.

In the *agnihotrabrāhmaṇa*s, karmic retribution is described in terms of a constant exchange between this earthly *loka* and that yonder *loka*. The offerings themselves construct (*sam+√kṛ*) the worlds as well as the body of the Agnihotrin. Ritual activity (*karma*) is also said to generate a something generated (*prajā́*) in yonder world, which at a later time is harnessed in *prāṇa* and takes embryonic form in the sense faculties of the Agnihotrin. In both the Brāhmaṇa texts and the Upaniṣads, the Agnihotrin's immortal body in the sun is beyond change for one who knows. Compare these almost identical passages from the *Taittirīya Brāhmaṇa* and *Bṛhadāraṇyaka Upaniṣad*:

> *eṣá nityó mahimā́ brāhmaṇásya | ná kármaṇā vardhate nó kánīyān |*
> *tásyaivā́tmā́ padavít tám̐ viditvā́ | ná kármaṇā lipyate pā́pakena |*
>
> (TB 3.12.9.7-8)

This is the eternal greatness of the brahmin. He does not increase by *kár-man*, nor does he become less. His *ātman* knows the path. Knowing him (the *ātman*) one is not polluted by evil *kárman*.

(tr. Bodewitz, 2019, 257; cf. Renou, 1952, p. 154)

> *ātmā́ ... sa ná sādhunā kármaṇā bhū́yán no evā́sā́dhunā kánīyān ...*
>
> *eṣa nítyo mahimā́ brāhmaṇásya na várdhate kármaṇā́ no kánīyān tásyaiva*
> *syā́t padavit tám̐ viditvā ná lipyate kármaṇā pā́pakenéti |*
>
> (BĀUK 4.4.22-23)

The *ātmán* is not greater by good karma, nor is it diminished by bad [karma] ... 'That eternal greatness of a brahmin is not increased or diminished by karma. That [*ātmán*] of him would be a knower of the path (*padavit*). Knowing this [*ātmán*] one is not stained (*√lip*) by bad karma'.

So far, there's continuity between the Brāhmaṇas and Upaniṣads.[53]

According to Herman Tull (1989, p. 105), the Upaniṣadic doctrine of karmic retribution grew out of Vedic ritual.[54] He asserts,

> The notion that certain acts lead an individual to attain a specific world in the afterlife is prefigured in the ritual sphere. This relationship between act and world is especially evident in a discussion of the results of the initiation, as the sacrificer—through the activity of the rite—is said to make (*/kṛ*) a world for his own self:

> When he performs the initiation, he makes for it [his own self] that world beforehand (*purastāt*); and when he becomes initiated he is born to that made

world. Therefore they say: 'Man is born to the world that is made [by his own self]'.

(ŚB 6.2.2.27)

But whereas in the Brāhmaṇa texts karmic retribution implicitly applies to all actions that take birth as *loka*s, depending on whether actions are performed while knowing in a certain way or not, in the Upaniṣads karma explicitly pertains to any action and leads to rebirth in accordance with the action with less optimism about knowing oneself. *Jaiminīya Brāhmaṇa* 1.2 and *Bṛhadāraṇyaka Upaniṣad* 4.4.5 go on to say:

amṛtā haivāsya prāṇā bhavanti | amṛtaśarīram idam kurute | so 'mṛtatvaṃ gacchati ya evaṃ vidvān agnihotram juhoti |

(JB 1.2)

His lifebreaths become immortal. He makes for himself an immortal body here (i.e. in the āgnyādhāna and agnihotra ritual). He becomes immortal who knowing thus offers the agnihotra (tr. Bodewitz).

yathākārī yathācārī tathā bhavati sādhukārī sādhúr bhavati pāpakārī pāpó bhavati púṇyaḥ púṇyena kármaṇā bhávati pāpaḥ pāpenátho khálv āhuḥ kāmamáya évāyaṃ púruṣa íti | sa yáthākāmo bhávati tátkratúr bhavati | yátkratúr bhavati tat kárma kurute | yat kárma kurute tád abhisáṃpadyate |

(BĀUK 4.4.5)

Whatever one does or whatever one practices, so he becomes. Doing good, he becomes good. Doing evil, he becomes evil. Meritorious by meritorious acts, evil by evil acts. Moreover, you should know, they say, 'A man is made of desire alone'. Whatever he desires, he becomes. That is his will (*kratu*). When there is will, then he creates karma. When he creates karma, then he is changed into that.

The dynamic exchange between the yajamāna's two bodies, the two worlds, a central element of the *yajñá* in the Brāhmaṇas, becomes replaced by the classical formula for karmic retribution – doing good he becomes good; doing evil he becomes evil – in the Upaniṣads.

In the late Vedic period, the need arose among specialists to present detailed sacrificial rules and procedures from the *apauruṣeya* Brāhmaṇas to students in a form that was easy to memorize.[55] The *pauruṣeya* Śrautasūtras codified and systematized Vedic doctrine precisely and briefly, presupposing a teacher who would provide interpretations and commentary (Gonda, 1977, pp. 465–466). Composed partly during and partly after the composition of the Yajurvedic Brāhmaṇa texts (471), the Śrautasūtras are based on and quote the injunctions of the Brāhmaṇa texts (486, 489, 497). Gonda (497) makes clear, the authors of the *brāhmaṇas* endeavoured to explain the origin, meaning and raison d'être of the ritual acts etc. and to prove their validity; the compilers of the *sūtras* (*sūtrakāra*), on the contrary, aimed at a systematic description of every ritual in its natural sequence.

In this way, the Śrautasūtras give directions for priests performing *śrauta* rituals, often omitting the motivations and explanations, for which one had to refer to a Brāhmaṇa text (498). Moreover, the sūtras prescribe more incidental rites and expiations than found in the Brāhmaṇa texts (499). A Śrautasūtra contains the injunctions (*vidhi*) of its corresponding Brāhmaṇa text, but almost no explanations (*arthavāda*).

After the Śrautasūtras were composed to study and carry out ritual practice, the Brāhmaṇa texts' cosmologies and explanations for rituals like the Agnihotra were no longer understood assiduously by all. Traditionally, there have been two types of Vedic scholars: one grounded in recitation, who learns the words by heart without necessarily knowing the meaning, and another grounded in interpretation.[56] As time went on, the Vedic scholars who recited without knowing the meaning relied on the Śrautasūtras to perform rituals and lost touch with the philosophy in the *arthavāda* part of the Brāhmaṇa texts. This resulted in Indian tradition to some extent disregarding the *arthavāda* of the Brāhmaṇa texts early on, such that when ideas were imported into other Indian schools of thought, they were no longer recognized as coming from the Brāhmaṇa texts. In some cases, the concepts evolved into new expressions and ideas. The ritual element championed by the Śrautasūtras – because Mīmāṃsā philosophy and the great commentator Sāyaṇa followed this Kalpavedāṅga – became the legacy imposed on the Brāhmaṇa texts, while its philosophy remained in the shadows. Still, those who could interpret the meaning of *śruti* were able to enliven its concepts in different times and places.

One destination to which the philosophy of the Brāhmaṇas travelled was the teachings of Gotama Buddha, who grew up in Kosala, steeped in eastern Vedic culture. Gotama was familiar with the Agnihotra, which he calls the best of *yaññas*, ritual offerings (Sn 568: *aggihuttamukhā yaññā*; cf. Vin I 245). Nevertheless, he says that attending to the Aggihutta will not purify a man who has not conquered doubt (Sn 249).[57] Whereas the philosophy of the Brāhmaṇa texts is generally optimistic and assumes as a norm that Agnihotrins, in this very life, will realize their true nature, the teachings of the Buddha take for granted that, due to karma that has existed since time without beginning, beings continue in *saṃsāra* for countless lives unless they wake up. In addition, by the time of Gotama Buddha, some key Brāhmaṇical concepts lost their edge and became intellectually viable, losing their power to point to something beyond language.

The Awakened One taught an entire discourse, sutta number seven in the *Suttanipāta*, to Kosalan brahmins who asked whether brahmins these days live according to the brāhmaṇical Dhamma of ancient brahmins (*porāṇānaṃ brāhmaṇānaṃ brāhmaṇadhamme*). The Buddha said no and then described how brahmins of old conducted themselves virtuously, including when they performed a *yañña* properly (Sn 295–296). He lamented that they took a change for the worse (Sn 299: *tesaṃ āsi*

Table 3.1 Concepts in the Brāhmaṇas as compared to the Upaniṣads

Brāhmaṇa texts	Upaniṣads
Prajāpati and *bráhman* as absolute	*bráhman* and *ātmán* as absolute
Agni and *prāṇāḥ* as sense faculties	*vijñāna* and *prāṇāḥ* as sense faculties
Two *ātmán*s: two bodies in this and yonder world	Two *ātmán*s: generally the physical body and the self in the heart, but with many variations
Cause and effect in terms of making *loka*s, the exchange between bodies/worlds, and generating *prajá* or merit	Cause and effect as karma, i.e., karmic retribution
nāma and *rūpa* as two *bráhman*s, the means by which *bráhman* can experience relativity and by which the worlds are made continuous	*nāmarūpa* as how this world manifested (*vyākriyata*, BĀUK 1.4.7), *satya* (1.6.3)

vipallāso) and became greedy and violent.[58] Nevertheless, many brahmin sacrificers approach the Buddha for instructions on the *yañña* and the Buddha teaches them. For example, he teaches the young brahmin Māgha that one should offer and while offering, one should purify his mind under all circumstances. The *yañña* is the support for the *yajamāna*, established in which one overcomes aversion.[59] However, the Bhagavan tells Puṇṇaka that, aspiring for a state of existence, brahmins performed *yañña*s (Sn 1044: *āsiṃsamānā Puṇṇaka itthabhāvaṃ ... yaññam akappayiṃsu*). Even though the performers of *yañña*s offered oblations, conditioned by greed and attached to passion for existence, they did not cross over birth and old age (Sn 1046). The Bhagavan tells Puṇṇaka that one crosses over birth and old age who is without hunger (*nirāso*, Sn 1054). Freiberger (1998) argues that in the Pāli canon, sacrifice has been reinterpreted by substituting certain elements, by ethnicization, and by spiritualization. One does not find a standard response to the question of sacrifice in the Buddha's teachings, but a range of teachings appropriate to various interlocutors. In this way, Gotama sometimes enlivened and sometimes counteracted numerous concepts germane to the Agnihotra, of which I will provide just a few examples.

First, the Brāhmaṇa texts use the metaphorical domain of human progeny to depict the cycle of cause and effect. The effects of action are described in terms of *prajá* – something that has been generated, offspring or a creature – that Agni conveys to the sun. The sun then takes the form of an embryo and takes birth in Agni, the Agnihotrin's physical body and breaths, i.e., sense faculties. Similar to the *prajá* taking embryonic form in the senses in the *agnihotrabrāhmaṇa*s, Buddhism employs the agricultural metaphorical domain of a seed and fruit to describe the same process. However, in Buddhism the effect immediately becomes a seed or karmic potential that is destined to ripen into a fruit in the future. Here we may note that the sun serving as a storehouse for the unmanifest karmic potentials seems likely to have inspired Yogācāra's *ālayavijñāna*.

Second, the Brāhmaṇa texts describe the Agnihotrin as having two bodies. The physical body, called *ātmán*, is comprised of breaths that convey the offerings between the worlds. The immortal body made of offerings and merit is variously described as *ātman* (ŚBK), *śarīra* (JB), and *aṅga* (ŚBM). Similarly, (Sanskrit) *ātmabhāva* and (Pāli) *attabhāva* occur in the sense of a body throughout Buddhist texts. Perhaps '*bhāva*' was added in Buddhist discourses to clarify that the body (*ātman*) is only a condition, the result of past karma, and nothing with which to identify. Through the Agnihotra, the Agnihotrin too was to realize his true identity, which included his unmanifest body, and to directly see the offering made between the two bodies, by which he would overcome repeated death. Overtime, it is likely that these concepts lost their philosophical edge in Indian culture, such that some sacrificers may have assumed that intellectually understanding one's identity with the sun was sufficient, without directly seeing the exchange between the worlds or knowing the original, vast, nondual mind. Perhaps to offset this complacency, Buddhadharma uses the word *kāya* to emphasize that the body is just a collection of many parts without any substantial core. In addition, the Buddha taught that all conditioned things are *anātman*, though *ātman* in the sense of *tathāgatagarbha* and the Buddha nature (*buddhadhātu*) within also occurs (Bhattacharya, 2015; Jones, 2021). Understood in the context of the philosophy of the Brāhmaṇa texts, Gotama's teachings can be seen as correcting some of his followers who had gone too far in one extreme without really knowing things as they are. Interestingly, a Buddha also comes to be described in terms of a body: the Dharmakāya, Sambhogakāya, Nirmāṇakāya. One who is awakened can manifest bodies, much like the Vedic primordial being, though these bodies are re-envisioned as rescuing beings from *saṃsāra*.

Third, when the *agnihotrabrāhmaṇa*s present cosmologies in which Prajāpati creates Agni first, Agni is described as an insatiably hungry consumer. His eating of other creatures enables dualistic perception, through incessant birth and death, and the continuity of relative existence. Not only are Agni's bodies (*tanū*) described as hunger, thirst, and decay (TĀ 4.22), but when Agni's many forms become thirsty, the one establishing the fires in the Agnyādheya also becomes thirsty. However, Agni's numerous bodies (*agnes tanvaḥ*) may be made free from thirst by offering a full spoon of clarified butter (Pūrṇāhuti), which causes the one establishing his sacred fires to become free from thirst (KS 8.3.12, MS 1.6.7; Navathe, 2012, p. 23; Krick, 1982, pp. 415-416). Hunger and thirst are key concepts in late Vedic philosophy. Buddhism mentions hunger, but focuses more on thirst (Sanskrit *tṛṣṇā*; Pāli *taṇhā*), as the driving force for grasping after karmic conditions to experience pleasure. Just as in the *agnihotrabrāhmaṇa*s Agni is hungry to consume the *prajá* from the moment he is born, so, in Buddhism, living beings thirst for pleasure and suffer due to their craving for sensual pleasure, existence, and non-existence.

Fourth, in the Agnihotra cosmologies, day and night revolving was understood as repeated death and what exhausts the merit (*sukṛtá*) of a man. The cycle of the sun represents a wheel of cause and effect connected to the sun becoming an embryo and being born in the fire. Day and night, called unsurpassable oceans (*samudra*), represent change and conditionality in Vedic tradition. In Buddhism, this cycle is given the name *saṃsāra*, which is still seen as an ocean to cross over.[60] This alternation of light and dark in the *agnihotrabhāhmaṇa*s is elaborated further in the doctrine of *pratītyasamutpāda* or dependent arising. Table 3.2 makes some connections between the 12 links and elements in the philosophy of the Brāhmaṇa texts.[61]

Fifth, in the *agnihotrabrāhmaṇa*s, *loka* refers to a sphere of experience, of which there are three: this world (= earth), intermediate space, and yonder world (= sky/heaven). The *prajá* in the sun, i.e., yonder world, are harnessed to the breaths by rays of light where, becoming embryos, they take birth in the fire (ŚBK 3.1.9.1). Agni is homologized with the breaths, which in the plural denote the sense faculties. There is a direct correlation between 'this world' and the creatures born in the sense faculties. The three sacred fires are the *āyatana*s of the devas (*Kāṭhaka Agnyādheyabrāhmaṇam* 7). In addition, the sun is the womb and abode (*āyatana*) of Agni (TB 3.9.21.2-3), while Agni is the abode (*āyatana*) of the devas, the divine powers that become sense faculties (MS 3.1.10). In Buddhism a similar connection between the world and the sense faculties is established. In a pair of verses found in both the *Suttanipāta* and the *Saṃyuttanikāya*, the Buddha is asked, 'In what has the world arisen' (*kismiṃ loko samuppanno*) and the answer is given, 'In six the world has arisen' (Sn 168-9; SN 1.7.10).[62] Here six refers to the six internal sense bases (*āyatana*). The correspondence between the worlds and the senses resembles the philosophy of the Brāhmaṇas, but in Buddhism the senses are restricted to six. In addition, the Buddhist verses add new elements: intimacy is formed with the six external sense bases and, grasping the six sense objects, the *loka* is disturbed.

Twice the *Saṃyuttnikāya* explains the origin of the world as dependent on the sense faculties and their objects, the coming together of which gives rise to consciousness, which gives rise to contact, which gives rise to feeling and the rest of the 12 links of dependent arising (SN 12:44 and 35:107).[63] The passing away of the world depends on the cessation of craving, which in domino effect causes the cessation of the other links up to old age and death and the whole mass of suffering. In contrast, the Brāhmaṇa texts present relative existence as the much-desired journey of the primordial being, who worked hard to engineer a way to make the *loka*s continuous, not to stop. At the time of the Brāhmaṇa texts, brahmins were still close enough to their origins to have the trip of one lifetime, but clearly, by the time of Śākyamuni, the journey through conditioned experience had become harder and harder to exit.

In the *agnihotrabrāhmaṇa*s, *loka* closely connects to *ātmán* (body). Two bodies are mentioned: the physical body and the undying body, correlating to this world and yonder world. The Agnihotrin forms (*saṃskaroti*) his own *ātmán*, his body, made of libations and merit in the yonder world. The use of *sam* + √*kṛ* in this context seems to have influenced the term *saṃskāra* in Buddhism (Jurewicz, 2016, pp. 372–373). In the Brāhmaṇas, the ritual offerings become his body. Only the one who knows is released from *punarmṛtyu* and enters the sun; for him, relative existence ends. At this point the journey into personal existence is over and the Agnihotrin returns to the impersonal ground of being, Prajāpati. The *Jaiminīya Brāhmaṇa* makes very clear that for this to happen one must not identify with any personal self. Buddhist teachings suggest that some confusion regarding what continues, whether a personal *ātmán* or an impersonal *ātmán*, developed over time. This is not surprising, given the variety of ways that *ātmán* is used in the Upaniṣads. Vestages of the Vedic connection between *loká* and *ātmán* appear in the *Alagaddūpamasutta* (MN 22), but the sutta corrects the mistaken idea – an idea that does not occur in the *agnihotrabrāhmaṇa*s – that any personal self continues forever. The phrase *so loko so attā* occurs five times in the sutta, where it is called the *diṭṭhiṭṭhānaṃ* ('standpoint for [wrong] views') for an ignorant person, who considers all of the aggregates to belong to him and to constitute a self. A well-taught disciple, in contrast, sees that the aggregates do not belong to him or constitute a self, and rejects the idea that the *loka* is the *attā*:

so loko so attā, so pecca bhavissāmi nicco dhuvo sassato avipariṇāmadhammo, sassatisamaṃ tath' eva ṭhassāmīti, tam-pi: n' etaṃ mama, n' eso 'ham-asmi, na mêso attā ti samanupassati. So evaṃ samanupassanto asati na paritassatīti |
(MN 22)

'"The *loka* is the *attā*. Having died, I will be permanent, constant, eternal, not subject to change. I will continue on forever". This too he regards, "This is not mine, I am not this, this is not my *attā*." Regarding in this way, he is not agitated about what does not exist'.

MN 22 effectively argues for *anātman* in order to avoid grasping the five aggregates and continuing in *saṃsāra*. It rejects taking a personal *ātman* as a *loka*. The sutta goes on to say that someone should also not think that he will be annihilated upon relinquishing all attachments, views, and conditioned things.[64] If the sutta suggests that something is not annihilated upon giving up conditioned existence, then what continues? Is it possible that the two doctrines are not completely opposed after all?

Sixth, both Vedic tradition and Buddhism emphasize knowing and distinguish two modes of knowing. Only the Agnihotrin who knows stops repeated death and enjoys the original mind out of which all the worlds arise. What matters is not knowing 'something', but knowing in a certain way (*evam*). The Brāhmaṇa texts emphasize a kind of knowing in which the *prajā* are seen, but not engaged in dualistic perception. This leads to expanding the mind to include *sarvam*. Similarly, Buddhist texts contrast conditioned knowing, in which the aggregates are grasped to cognize a sense object, with unconditioned knowing, which is a direct knowing unmediated by and not dependent on any karmic conditions. Unconditioned

Table 3.2 Concepts from the Brāhmaṇas as compared to *pratītyasamutpāda*

Brāhmaṇa Texts	**Buddhist Dependent Arising**
darkness, not knowing in a certain way	1. *avidyā*
ātmānaṃ sam+√kṛ: the *yajamāna* makes (*saṃskaroti*) his body in yonder world, made of offerings and merit	2. *saṃskāra*
Agni	3. *vijñāna*
nāma rūpa = speech and mind	4. *nāmarūpa* = five aggregates
devāḥ prāṇāḥ → *deva*s take their place (*āyatana*) in the body, in the sense faculties	5. *ṣaḍāyatana*
	6. *sparśa*
	7. *vedanā*
hunger and thirst	8. *tṛṣṇā*
punarmṛtyu	9. *upādāna*
birth of the *prajā*	10. *bhava*
the son becomes the father	11. *jāti*
complete life = 100 years	12. *jarāmaraṇa*

knowing knows as it really is (*yathābhūtam*). In this way, the Vedic *ya evaṃ veda* (one who knows in this way) conveys an analogous thrust to the Buddhist *yathābhūtaṃ prajānāti* (one directly knows as it really is). Without dependence on past karma or dualistic intellectual cognition, the mind simultaneously, and in a non-dual way, knows, and is full of, everything.

Seventh, in the *agnihotrabrāhmaṇa*s, the Agnihotrin was to look down and see the revolving of day and night as one seated in a chariot would look down and see the wheels turn. Only then would his merit not be exhausted. Similarly, he was to sit between the fires, which represent the two worlds, and go to heaven where he gained the perspective of the offerings being exchanged before the *prajā* from yonder world take birth in his sense faculties. The idea of looking down is personified in the bodhisattva Avalokiteśvara, whose name literally means one who has the capacity of looking down. In iconography, he often has eyes in his thousand hands. Because he is able to see the conditions of all beings, he can respond effectively to remove their *duḥkha* and bring them closer to awakening. The *yajamāna* in the Brāhmaṇas looks down from *svarga* to see the exchange between the two worlds, while the bodhisattva looks down to see the conditions of all living beings.

Eighth, an optional practice in the Agnihotra, particularly after the evening libations are offered, consists of standing near the two fires and worshipping the fires by means of the recitation of various mantras. The root *upa+√sthā* literally means to stand close, to be present, attend to, or worship. The two fires represent the two worlds and standing between them prompted an awareness of the offerings moving back and forth, whether physically represented as pouring libations in the fire or mentally as the in and out breaths feeding the two bodies. Is it possible that the Agnyupasthāna inspired *smṛtyupasthāna*? Of course, in Buddhism, establishing mindfulness is divided into body, feelings, mind, and dharmas, categories which are not specified in Vedic thought. But perhaps the basic idea behind *smṛtyupasthāna* was taken from this Vedic ritual practice.

Nineth, in the Agnihotra cosmologies, the primordial being, sometimes homologized with *manas*, desires to generate himself or have a companion. Desire for something initiates dualistic perception and the creation of creatures with constant *punarmṛtyu*. Similarly, in the *Śūraṅgama Sūtra*, when the mind moves, understanding is added to *bodhi*, together with something to be understood (145). From this comes a mental strain on the awakened mind, giving rise to the sense faculties and the grasping at objects. Thus, motion becomes the world of what is to be known. From adding understanding to *bodhi*, which creates a disturbance in the mind, the primary elements arise along with mountains, rivers, faculties, beings, etc. (146). Adding understanding to awakening leads to the original mind becoming entangled in knots. The cosmology found in the *Śūraṅgama Sūtra* follows many cosmologies found in the Brāhmaṇa texts.

Vedic scholar K.R. Potadar (1953, p. 284) once said, 'Ideas are like seeds, sprouting in different regions at different times'. The philosophies of the Brāhmaṇa texts and Buddhism travelled in the *Bhagavadgītā*, where the *yajña* is recast in light of devotion to Kṛṣṇa. Like the *agnihotrabhrāhmaṇa*s, the *Gītā* speaks of two bodies. However, Kṛṣṇa distinguishes the bodies that come to an end (*antavanta*

ime dehāḥ) with the eternal, embodied, indestructible, immeasurable (*nityasya...
śarīriṇaḥ anāśino 'prameyasya*)[*ātman*] (BhG 2.18). Kṛṣṇa calls on his faithful to
literally bring into being, i.e., nourish, the devas (3.11: *devān bhāvayata*) so they
bring us into being (*te devā bhāvayantu vaḥ*). The devas are described as brought
into being or nourished by the *yajña* (3.12: *yajñabhāvitāḥ*). Like in the Vedas,
the goal is to see the various states of being situated in the one (*ekastha*) so as to
attain *brahman*.[65] Still the primordial being, as in the *agnihotrabrāhmaṇa*s, *brahman* in the *Gītā* is seen in light of Kṛṣṇa, the imperishable, and the unmanifest.[66]
The *Gītā*'s understanding of the *yajña* follows the same logic. The *yajña* is one
whose origin is *karma* (3.14) and *karma* arises from *brahman* (3.15), which is forever established in the *yajña*. Every aspect of the *yajña* is *brahman*, including the
offering poured by *brahman* into the fire of *brahman* (4.24). Similarly, Kṛṣṇa is all
aspects of the sacrifice: the *yajña*, the mantra, the ghee, the fire, and the act of offering (9.16).[67] Whatever is offered is offered to Kṛṣṇa, who is the ultimate recipient of
all sacrifices (9.24, 27). Just as in the *agnihotrabrāhmaṇa*s Agni is the womb into
which the embryo was placed, Kṛṣṇa declares that *brahman* is his womb, in which
he places the embryo, and all creatures (*bhūta*) come from that (14.3).[68]

The Brāhmaṇa texts describe the Agnihotra as an offering between the worlds,
in which the unmanifest yonder world offers into Agni, who is the breaths and the
sense faculties. The *Gītā* describes how some offer the sense organs in the fires
that are self-restraint and the sense objects into the fires that are the senses (4.26).[69]
Others offer the *karma*s of the senses and breaths into the fire that is self-restraint
illumined by knowledge (4.27).[70] Still others offer with material things, austerity,
yoga, study, and knowledge (4.28).[71] Some even offer breath in breaths; Kṛṣṇa says
that all of them are knowers of the *yajña* who have their imperfections obliterated
through *yajña*.[72] Nevertheless, Kṛṣṇa says that the *jñānayajña* is better than a *yajña*
consisting of material things (4.33: *śreyān dravyamayād yajñāj jñānayajñaḥ*).

The *yajña* is still seen as central to religious practice in the *Gītā*, but Kṛṣṇa specifies a particular way to perform sacrificial action. One should be free from attachment (3.9: *muktasaṅga*). The karma of *yajña* should be undertaken, not abandoned
(18.5), but only insofar as one abandons attachment (*saṅga*) and the fruits (*phala*)
of karma (18.6). Such a one would be *sthitaprajña* (2.54-55), *sthitadhīḥ* (2.54), and
tasya prajñā pratiṣṭhitā (2.58, 61, 68).

In the *Bhagavadgītā*, new elements from Buddhism, Sāṃkhya, and Yoga are
interwoven with the ancient Vedic teachings, which are now seen as superfluous
for a brahmin who knows (2.46: *brāhmaṇasya vijānataḥ*).[73] Like Buddhadharma,
the *Gītā* mentions contact (2.14), abandoning desire (2.71), and *saṃsāra* (9.3), as
well as *brahmanirvāṇa* (2.72). Sacrificers who reach heaven are said to be reborn
in the mortal world (9.20-21), in contrast to the Vedic understanding that reaching
heaven implied becoming whole in an impersonal way after directly knowing the
unmanifest, immortal body. The *Bhagavadgītā* weaves in elements of Sāṃkhya
philosophy, such as the *guṇa*s, and Yoga, which it describes as evenness (2.48:
samatvaṃ yoga ucyate) and skill in action (2.50: *yogaḥ karmasu kauśalam*). Ideas
from many Indian systems of thought travelled in the *Bhagavadgītā*, including
ideas from the *agnihotrabrāhmaṇa*s.

Śaṅkarācārya wrote a detailed commentary on the *Bhagavadgītā*, in which he quotes the '*śātapathīya brāhmaṇa*' – meaning the *Bṛhadāraṇyaka Upaniṣad* part – and mentions *śrauta* rituals like the Agnihotra.[74] He understands the Agnihotra and other ritual karma to be performed by those who seek desires, beginning with heaven, and considers sages of old, like Janaka, to be knowers of reality (*tattvavid*) who performed ritual karma for the sake of others or if not knowers of reality, they remained established in perfection (*saṃsiddhi*).[75] In his introductory remarks to chapter 3, Śaṅkarācārya addresses the question of *nityakarma* (obligatory karma), asserting that such rites do not apply to a *sannyāsin*. The Agnihotra is a *nitya śrauta* ritual, so performing it twice daily is obligatory for any Āhitāgni. The Agnihotra can function as a *kāmya* (based upon desires, i.e., optional) ritual by substituting the oblation substance and mantras. However, a *kāmya* Agnihotra would only be performed once or twice to fulfill specific desires. Śaṅkara's comments about performing the Agnihotra with the desire for heaven are misleading from the perspective of the Brāhmaṇa texts, insofar as the Agnihotra is primarily a *nityakarma*, which is done without expecting any result. Moreover, properly done, the Agnihotra leads the Agnihotrin to heaven with every performance, without any special desire for heaven. Breaking with the Brāhmaṇa texts, Śaṅkarācārya refutes the idea that not performing *nityakarma* causes offense (*pratyavāya*). The idea of a renunciant who cultivates liberation through the mind influences his reading of Vedic tradition.

The Advaitin philosopher advocates a stark division between *karma* and *jñāna* in his commentary on *Bhagavadgītā* 2.10 and the first four verses of chapter 3. These sections forge a separation between adherence to *jñāna* or *karma*, which contrasts sharply with the inseparability of knowing and ritual action in the Brāhmaṇa texts. Śaṅkarācārya concludes his remarks on verse 2.10 by saying, 'Therefore, in the *Gītā* treatise, attaining liberation is exclusively through knowing reality, which is not combined with ritual action'.[76] At the end of his introduction to chapter 3, he speaks of, 'A failure of proof regarding the combination of *jñāna* and *karma*. Therefore, 'liberation is exclusively from knowing' is the meaning ascertained in the *Gītā* and all the Upaniṣads'.[77] Similarly, at the end of his commentary to verse 3.3, he states, 'Therefore, by no reasoning whatsoever is there a combination of *jñāna* and *karma*'.[78] In addition, these two kinds of yoga are to be performed by different people – adherence to *jñāna* by *sannyāsins* alone.[79]

Śaṅkarācārya explains (on 2.10) that when the discriminating consciousness is overwhelmed by grief and confusion (*śokamohābhyāṃ hy abhibhūtavivekav ijñānaḥ*), grief and confusion become the seeds of *saṃsāra* (*saṃsārabījabhūtau śokamohau*). This is a more abstract and elaborate expression of the mechanism, found in the Brāhmaṇa texts, of the sun as an embryo or *prajā* taking birth in Agni. So, while performing action without attachment (BhG 5.11) is an important means of purification (18.5), freedom ultimately comes from *jñāna*. Whereas, in the Brāhmaṇa texts, performing ritual expands the mind and leads to the highest goal, for Śaṅkarācārya, *karmayoga* is only a means to *jñānayoga*.[80] Karma is a preliminary step to purifying the mind, and one should perform karma. However, ritual karma is not the ultimate aim for Vedānta philosophy, but only a means to it.[81]

Table 3.3 Concepts in the Brāhmaṇas as compared to those of the Upaniṣads and Buddhism

Brāhmaṇa texts	Upaniṣads	Buddhism
Prajāpati (= *manas*) and *bráhman* as absolute	*bráhman* and *ātmán* as absolute	*prajñā, tathāgatagarbha, buddhadhātu, ātman Dharmakāya*
Agni and *prāṇāḥ* as sense faculties devas take their places (*āyatana*) in the body *prāṇāḥ* = sense faculties, to which are harnessed *prajā́* born in 'this world'	*vijñāna* and *prāṇāḥ* as sense faculties	*ṣaḍāyatana*s the origin of the world = *ṣaḍāyatana* and *pratītyasamutpāda* starting with the sense faculty meeting the sense object
two *ātmán*s or bodies in this and yonder worlds	two *ātmán*s: body and Self/*bráhman* in the heart	body as *ātmabhāva* and *kāya*; *loka* as *ātman*; *anātman* and *ātman*
cause and effect in terms of making *loka*s, the exchange between bodies/ worlds, and generating *prajā́*	cause and effect as karma	cause and effect as karma
nāma and *rūpa* as two *bráhman*s, the means by which *bráhman* can experience relativity and by which the worlds are made continuous	*nāmarūpa* as how this world manifested (*vyākriyata*, BĀUK 1.4.7), *satya* (1.6.3)	*nāmarūpa* as the 5 *skandha*s and as one of the 12 links of *pratītyasamutpāda* (giving rise to the *āyatana*s)
hunger, thirst	hunger	thirst
Agnyupasthāna		*smṛtyupasthāna*
day and night revolving		*saṃsāra*
ya evaṃ veda	*ya evaṃ veda*	*yathābhūtaṃ prajānāti*
looking down at the revolving of day and night		Avalokiteśvara
sun as the receptacle of offerings and merit (*sukṛtá*)	sun, heart	*ālayvijñāna*
cosmologies starting with desire arising in the primordial being = mind	cosmologies starting with desire arising in the primordial being	cosmology starting with the movement of the mind

Like Vedānta, Mīmāṃsā divides the Vedas into the Karmakāṇḍa, which deals with the ritual offerings from the mantra and Brāhmaṇa portion, and the Jñānakāṇḍa, which focuses on knowing *brahman* as described in the Upaniṣads (Sharma, 1987, p. 211). The former is the focus of Pūrva-Mīmāṃsā, while the latter that of Uttara-Mīmāṃsā or Vedānta. Of course, there were teachers like Kumārila and Maṇḍana Miśra against whom Śaṅkarācārya argues, who advocated the combination of *karma* and *jñāna* (*karmajñānasamuccayavāda*). Such proponents continue the position of the Brāhmaṇa texts, in which *karma* and *jñāna* formed an inseparable pair, such that one should perform a ritual karma *ya evaṃ veda*.

Like the Śrautasūtras, Mīmāṃsā is mainly concerned with *vidhi* (injunctions) from the Brāhmaṇa texts, deeming the *arthavāda* portions to be secondary, subordinate, and supplementary to injunctions. Dharma is understood as what is indicated by injunctions (*Śābarabhāṣya* 1.5.3; Jha, 1933, p. 25). The explanations are not only to be construed with injunctions but indicate meaning only when syntactically connected to injunctions (JMS 1.2.22–25). In addition, Śabara maintains that the explanations contradict directly perceived facts, giving as an example from the *agnihotrabrāhmaṇa*s that the fire from this world enters the sun and at night the sun enters the fire. The *siddhānta* asserts that *arthavāda* sentences unconnected to action, when they construe with injunctions, serve as a valid means of knowledge (ŚBh 1.2.7; Jha, 1933, p. 55). He allows for indirect or figurative expression for the sake of praise and the mention of events to arouse attraction so that one undertakes an act, or repulsion, so one desists from an act (1.2.10; Jha, 1933, p. 58–60). To preserve the timelessness of the Veda, however, Śabara regards the mythological stories of the Veda as allegorical (1.2.10; Verpoorten, 1987/2020, p. 14). Mīmāṃsā's take on myths in the *arthavāda* portion of the Brāhmaṇa texts marks a departure from the Brāhmaṇa texts, for which the cosmologies and explanations of the rituals were inseparable from injunctions and indispensable to understanding the connection between humanity, ritual action, and *sarvam*.

Earlier Mīmāṃsakas advocated Dharma and aimed to attain *svarga* (heaven), but later Mīmāṃsakas put their faith in *mokṣa* and replaced the aim of attaining *svarga* with liberation (*apavarga*) (Sharma, 1987, pp. 212, 237). For Mīmāṃsā, one performs rites to achieve the bliss (*prīti*) called *svarga* (Verpoorten, 1987/2020, p. 19). This restyling of *svarga* as that which is full of pleasure, without any pain, reflects a later development in Indian thought because, according to the Brāhmaṇa texts, attaining heaven was connected with a knowing mind capable of looking down to directly see the exchange between the unmanifest yonder world and this world. Unlike in the philosophy of the Brāhmaṇas, where reaching *svarga* reintegrates the mind's innate wholeness, the desire for *svarga* and pursuit of liberation as understood in Mīmāṃsā bifurcates the mind. Mīmāṃsā's term for liberation (*apavarga*) comes from the root *apa*+√*vrj*, which means to exclude. In the context of Mīmāṃsā, liberation referred to separation from worldly existence, exemption from the cycle of birth and death and from pain. There is no concept of liberation as exclusion in the Brāhmaṇa texts, for which the goal was to integrate the two worlds back into one whole, referred to as *sarvam*. Mīmāṃsā claims to follow and properly interpret the Veda. In reality, however, Mīmāṃsakas use the authority of the Veda to advance their own ideas.

Prabhākara and Kumārila regard the *ātman* as the eternal and omnipresent knower, enjoyer, and agent that is different from the body. This conception is at odds with the doctrine of the two bodies advanced in the Brāhmaṇa texts and reflects, instead, Upaniṣadic developments (Sharma, 1987, p. 234; cf.

Verpoorten, 1987/2020, p. 19). Whereas the *agnihotrabrāhmaṇa*s understand the sun to be both an extension of oneself and the receptacle of one's embryonic creatures created through action, Mīmāṃsā advances *apūrva* as an unseen potency, produced through actions, in the Self. In this way, Mīmāṃsā reinterpreted both *ātman* and cause and effect when the philosophy of the Brāhmaṇas travelled in time and place.

Concerned with the interpretation of Vedic texts, Mīmāṃsā provided philosophical justification and rational arguments for Vedic views. Whereas Śabara enjoins the enquiry into Dharma only after reading the Veda and directly quotes from the Vedas, later Mīmāṃsakas rely on Śabara's Vedic quotations. Mīmāṃsakas introduce a system of *pramāṇa*s, originating outside the scope of the Brāhmaṇa texts, to codify how to interpret the Vedas. The *apauruṣeya* Brāhmaṇa texts are rooted in a type of knowing that goes beyond intellectual thinking, whereas in Mīmāṃsā, intellectual knowing and reasoning is fundamental to many of the *pramāṇa*s. Mīmāṃsakas claim to be authorities on the Veda, but they do not represent the Brāhmaṇa texts in their entirety.

Reflecting on how mobilities of ritual in the *agnihotrabrāhmaṇa*s travelled throughout Indian intellectual history, it is clear that already during the period in which the Brāhmaṇa texts were composed, the Śrautasūtras championed the injunctions to perform Vedic ritual, while rendering meaningless independent *arthavāda* portions of the Brāhmaṇa texts not construed with injunctions. This split proved to have profound consequences for the subsequent understanding and transmission of Brāhmaṇical philosophy. Similarly, Mīmāṃsā relegated the *arthavāda* to subsidiary importance, useful and meaningful only when connected to an injunction. The Upaniṣads and Buddhism refashioned the philosophy of the Brāhmaṇas, keeping some basic concepts while enlivening others and adding additional ones. Buddhism in particular used some contrasting concepts in an attempt to bring cultivators back to the middle and out of *saṃsāra*, but its foundation is rooted in Vedic thought. The *Bhagavadgītā* recasts aspects of the *yajña* in terms of a personal absolute, Kṛṣṇa, and specifies that sacrificial action should be performed without attaching to the fruit of karma. Śaṅkarācārya's Vedāntic commentary on the *Gītā* maintains that insofar as ritual action (*karma*) purifies, it is a means to *jñāna*. However, liberation comes only from knowing. Similarly, Mīmāṃsā divided the Vedas into a Karmakāṇḍa and a Jñānakāṇḍa. The division gives the false impression that the Brāhmaṇas and the Upaniṣads are separate philosophies. Given the lack of attention to the *arthavāda* portion of the Brāhmaṇa texts since the inception of the Śrautasūtras and the dominant ritual hermeneutics that ensued as a result, many throughout history never questioned this division. However, *karma* vs. *jñāna* is a false dichotomy: both were equally important and inseparable in the Brāhmaṇa texts. In addition, Mīmāṃsakas redefined *svarga*, adopted the Vedāntic notion of *ātman*, restyled liberation, and added *apūrva* and a new methodology for Vedic interpretation. In this way, the philosophy of the Brāhmaṇas traveled quite a distance in Indian intellectual history. Its thought became both enlivened in new contexts and silenced by the clamour of new intellectual developments and praxis.

Vedic Scholarship during Colonialism and Beyond

The treatment of the Agnihotra in later Vedic tradition and philosophical schools like Vedānta and Mīmāṃsā contributed to the philosophy of the Brāhmaṇas having a rough trip during the colonial period and beyond. E. Windisch, Charles Schwab, Thomas Trautmann, and Michael Dodson sketch the initial reception of the Vedas during the colonial period. The sixteenth century missionaries in India relied on oral information from local brahmin pandits. It was not easy and took them a long time to learn Sanskrit. In addition, much of what they were initially told about the Vedas was inaccurate. For example, with the help of Brāhmaṇa Padmanāba, the Dutch missionary Abraham Rogers wrote a chapter on 'Vedam' that purports to describe the contents of the four Vedas, but his description is actually based on Tamil Vaiṣṇava hymns (Windisch, 2008, p. 3). Early books claiming to describe Vedic tradition had little content from the Saṃhitās and Brāhmaṇas, and instead related information from post-Vedic, especially Purāṇic, tradition (Ibid., pp. 13–14). Eighteenth century letters written by Jesuits indicate that they believed India was indebted to the West for its knowledge, assuming that the Vedas had been copied 'from the laws of Moses' (Schwab 1984, p. 146). A letter in 1735 by Father Calmette mentions some philosophical teachings of the 'Vedam' and another two years later indicates that the king instructed him to acquire Vedic texts (Windisch, 2008, p. 11).

The early orientalist works in the 1760s rely on Persian translations or vernacular renderings of Sanskrit texts (Dodson 2007, p. 43). The first Sanskrit grammars were written in Latin in the eighteenth century and printed in Rome in 1790 and 1804 (Schwab, 1984, p. 32; Windisch, 2008, p. 33). Schwab explains that 'biblical antiquity had been the most ancient past known; henceforth its priority would be a matter of contention' (Schwab, 1984, p. 5). Trautmann comments, 'Hinduism, in Holwell, reads like Milton with transmigration' (Trautmann, 1997, p. 69). In 1765, Holwell suggested that the early biblical accounts may be incomplete and should incorporate earlier Hindu stories.[82] A great admirer of Hinduism, Trautmann says he basically rewrote Christianity using Hinduism, calling for the reform of the church. In response, Christianity turned to rationality and science to protect the legitimacy of its claim to truth (Dodson, 2007, p. 29). Like the Church, the East India Company had an interest in Sanskrit texts.

Company employees Sir Charles Wilkins, Sir Williams Jones, and Henry Thomas Colebrooke made progress in learning and translating Sanskrit in the late eighteenth and nineteenth centuries. Son of the chairman of the East India Company, Colebrooke published 'On the Vedas, or Sacred Writings of the Hindus' in 1805 [1873]. In this essay Colebrooke says that the Vedas are too voluminous to be translated and hardly worth the trouble to the reader or to the translator (Windisch, 2008: 49).[83] This essay conveys the knowledge and interests of the Indian pandits of his time. Relying on actual manuscripts of the *Ṛgveda*, Colebrooke discusses the hymns and describes the other three Vedas. His mention of the *Aitareya*, *Śāṅkhāyana*, and *Śatapatha Brāhmaṇa*s is the first by any European, though the Brāhmaṇa texts did not seem to be of interest to the pandits of north India at that time (Ibid., p. 122). He also writes on the Upaniṣads, emphasizes the *Nirukta*, and cursorily mentions Sūtra literature. From the very first serious study of the Vedas, before any edition or any full translation of any Vedic text had been made, the opinion formed that translating them would be a waste of time.

Despite this, Colebrooke's colleague, Horace Hayman Wilson's last major work was to translate the *Ṛgveda* in six volumes (1850–1888), which coincided with the edition of the *Rig-Veda-samhitâ* with Sāyaṇa's commentary by F. Max Müller published in six volumes between 1849 and 1874 (Ibid., p. 80). Müller was just 26 years old when the first volume was published (Ibid., p. 478). The East India Company supported the printing of both Wilson's and Müller's work on the *Ṛgveda*.[84] Note that Sāyaṇa is well versed in Mīmāṃsā (Ibid., p. 479). Friedrich Rosen, who studied under Bopp, translated into Latin the first *aṣṭaka* of the *Ṛgveda* on the basis of two manuscripts in 1830 (Ibid., p. 163). His work was published posthumously in 1838. J. Stevenson made an edition of the first 39 hymns in 1833. Allexandre Langois, a student of Chézy, translated the *Ṛgveda* into French between 1840 and 1851, before a complete edition of the text was available. However, Müller, Otto Böhtlingk, and Rudolph Roth commented that his grammar was not very exact (Ibid., p. 247). German translations were published by Hermann Grassmann in 1876–1877 and by Alfred Ludwig in 1876. Theodor Aufrecht edited the *Ṛgveda* in Roman with only the root text in 1877. Hermann Oldenberg published *Vedic Hymns* starting in 1891. Ralph Griffiths's English translation came out a few years later. By the end of the nineteenth century, editions of the *Ṛgveda* and English, French, and German translations were available.

Along with these works on the *Ṛgveda* came the study and translation of the Upaniṣads. Anquetil Duperron translated the Upaniṣads in 1801 from the Persian version by Dara Shikoh (son of Shah Jahan) into French and Latin (*Oupnek'hat*) (Schwab, 1984, pp. 52, 66; Windisch, 2008, p. 87). Ram Mohun Roy, founder of the Brahmasamāj, also published translations of Upaniṣads in 1816 and 1819. Rejecting all rituals and advancing the Upaniṣads and Vedānta, Roy commenced the first edition of the *Vedāntasūtra* with Śaṅkara's commentary. Böhtlingk published a translation of the *Chāndogya Upaniṣad* and *Bṛhadāraṇyaka Upaniṣad* in 1889. In 1897, Paul Deussen translated fifty Upaniṣads based on Sanskrit texts. With these works translated first, it is no surprise that intellectuals in Europe had a slanted view of Vedic tradition. For example, Friedrich Schlegel believed that the Upaniṣads were the proper basis of Indian theology (Windisch, 2008, p. 105). His brother, A.W. Schlegel, despite having only read the Upaniṣads and not the *Ṛgveda* or the Brāhmaṇa texts, opined that 'no progress' had been made from the sensuous toward the spiritual, but rather the reverse (Ibid., p. 142). Louis Poley published the Calcutta edition of a few Upaniṣads with Sāyaṇa's commentary and translated a couple of them into French in 1844 (Ibid., p. 165). He published a translation of 20 hymns from the *Ṛgveda* from Rosen's edition, and three more Upaniṣads, but died before he could finish. The publisher published the unfinished book with a note about why the book was never finished.

Following Stevenson's problematic edition and translation of the *Sāmaveda* in 1842 and 1843, Theodor Benfey published a German edition in 1848, though he did not have access to Sāyaṇas commentary on the *Ṛgveda* for the first part (Ibid., pp. 393–394). Ralph Griffith published an English translation in 1893. W.D. Whitney and Roth edited *Atharva Veda Saṃhitā* in 1856. After studying classical

Oriental philology for only three years and completing his dissertation on the ninth Adhyāya of the *Vājasaneyi Saṁhitā* with Mahīdhara's commentary, Albrecht Weber edited the *Vājasaneyi Saṁhitā*, titled *White Yajurveda* (1852). Griffiths published a translation of the *Yajurveda* in 1899. Leopold von Schroeder edited the *Maitrāyaṇī Saṁhitā* (1881–1886) and the *Kāṭhaka Saṁhitā* (1900ff). Weber edited the *Taittirīya Saṁhitā* in 1871 and 1872 and A.B. Keith translated it in 1914.

The Brāhmaṇa texts were studied comparatively less and slightly later compared to the *Ṛgveda* and Upaniṣads. Weber edited the *Mādhyandina Śatapatha Brāhmaṇa* (1855) and the *Śrautasūtra of Kātyāyana* (1852, 1855, 1859). He was the first to publish an edition of a Brāhmaṇa. The three volumes were printed under the patronage of the East India Company. Martin Haug edited the *Aitareya Brāhmaṇa* in 1863 and Aufrecht edited it again in 1879. Julius Eggeling published a translation of the *Mādhyandina Śatapatha Brāhmaṇa* in multiple volumes (1882, 1885, 1894, 1897, 1900). Hanns Oertel published an edition and translation of the *Jaiminīya or Talavakāra Upaniṣad Brāhmaṇa* in 1894. By the end of the nineteenth century, there were only critical editions of a few Brāhmaṇa texts and a couple of translations.

Secondary literature claiming to represent early Indian thought was also published during the colonial period. Under the spell of Adam Smith's 'progress', before any edition or translation of the Vedas was available, and without reading any primary sources directly, James Mill rejected the idea that ancient Indian society was an advanced civilization (Dodson, 2007, p. 66). In his 1817 book *The History of British India*, Mill defended his lack of knowledge of Indian languages or the fact that he had never travelled to India by claiming that his use of mental faculties to judge testimonies was more important than gathering facts with the senses in the field. According to Trautmann, Mill's 'intent to minimize the Indian accomplishment and to subvert the expertise of the Orientalist is evident on every page' and demonstrates 'that theory offers inlets for prejudice the equal, at least, of impressions of sense' (Trautmann, 1997, p. 121). His imperfect history was heavily criticized by H.H. Wilson, who went so far as to say that 'its tendency is evil' (Ibid., p. 118). Wilson edited later editions of Mill's work with notes based on his reading of the texts and extended the period of study under the title *History of British India from 1805 to 1835* (1844–1848). Nonetheless, Mill's book was required reading for the Company's civil servants and was read by notable intellectuals.

For example, in 1822–1823, Georg Wilhelm Friedrich Hegel (1770–1831) taught a series of lectures that sought to bring all of world history into his dialectic of the spirit. He based his notions of India on Mill's account, colonial reports, and journal articles circulating in Europe. His lecture notes were published posthumously. Hegel decided that in India, imagination dominates over reason. In his words, India:

has always been the land of imaginative aspiration, and appears to us still as a Fairy region, an enchanted World...India is the region of phantasy and sensibility...This idealism, then, is found in India, but only as an Idealism of imagination, without distinct conceptions; – one which does indeed free existence from Beginning and Matter [liberates it from temporal limitations and gross materiality], but changes everything into the merely Imaginative; for although the latter appears interwoven with definite conceptions and

Thought presents itself as an occasional concomitant, this happens only through accidental combination (Hegel, 1956, p. 139).

Like Kant, Hegel presents the imagination as the most base faculty, under the faculty of understanding, which is inferior to reason. In adhering to values of European rationalism, Hegel is unable to understand the philosophy of India as anything but stuck in the most inferior level of the faculties. He sees Indian thought as 'rich in imagination and genius' and yet 'unworthy' in every respect (Ibid., p. 140). For him, the Indian view, full of symbols and mythology, is 'stripped of rationality' (Ibid., p. 141). Hegel fell prey to the limited understanding of India available to him according to a colonial episteme that privileged reason.

From 1818 to 1829, Hegel taught a series of lectures on aesthetics, which were compiled by one of his students in 1835. In these lectures, he situates India within an early period of development, pre-Classical Greek art, in which myths were expressed not through thought but imagination. Hegel writes,

Indian art, although severing universality from individual existence, nevertheless demands the immediate unity of both as well, a unity produced by the imagination [*Phantasie*]; it must therefore deprive the determinate existent of its limitlessness and, in a purely sensuous way, enlarge it into indefiniteness and, in general, transform and disfigure it (Hegel, 1975, p. 339).

Without determinacy, the limited shape expressing the absolute gives rise to what appears sublime, but is not sublime, symbolical, or beautiful. Hegel maintains,

Indian art contains many things which begin to strike this note of sublimity. Yet the great difference from sublimity, properly so-called, consists in this, that the Indian imagination [*Phantasie*] in such wild configurations does not succeed in positing negatively the phenomena that it presents, but precisely by that immeasurability and unlimitedness thinks that the difference and contradiction between the Absolute and its configuration has been obliterated and made to vanish (Ibid., p. 340).

Hegel cannot accept the Indian configuration because he sees it as degrading the external shape that bears a universal meaning, expressed in his terms as the 'one God' who is 'explicitly without shape and is incapable of expression in his positive essence in anything finite and mundane' (Ibid., p. 372). What he writes about India says more about his penchant for a rational ideal than anything pertaining to Indian tradition: hardly any translations of the Vedas were available during his lifetime. Nonetheless, Hegel's criticism of mythological symbolism likely discouraged orientalists from studying the *arthavāda* of the Brāhmaṇa texts, which is rich in myths and metaphors.

Most Indological scholarship on the Vedas that was grounded in a better, though ever evolving, understanding of Vedic texts came after Hegel. Wilson, with the help of pandits, published his essay 'A sketch of the religious sects of the Hindus' in 1828 and 1832. When P.V. Bohlen published *Das alte Indien* in 1830, no translation or edition of the Vedas was available, but when Émil Burnouf, nephew of Eugène Burnouf, wrote *Essai sur le Veda ou Études sur les religions, la literature, et la constitution sociale de l'Inde despuis les temps primitifs, jusqu' aux temps brahmanique* in Paris in 1863, the *Ṛgveda* was known through the translations by Rose, Langlois, and Wilson, in addition to Max Müller's edition (Windisch, 2008, p. 257). Benfey published a long encyclopedia article 'Indien' in 1840, for which Mill's *History of British India* was a source (Ibid., p. 282). Seven years later, Lassen published the first volume of *Indische Alterthumskunde* (Indian Archaeosophy) in Bonn.[85] The Brāhmaṇa texts were not sufficiently known to him in the first edition, in which he only quotes from the *Aitareya Brāhmaṇa*. In the second edition he also mentions the story of Māthava Videgha from the *Śatapatha Brāhmaṇa*, which Weber had discussed (Windisch, 2008, p. 293). He draws from Roth's 'Zur Litteratur und Geschichte des Weda' and, in the second edition, also from Müller's *History of Ancient Sanskrit Literature* (1859), Weber's *White Yajurveda* (1852), *Śatapatha Brāhmaṇa* (1855), and Roth and Whitney's *Atharvaveda* (1856). When he was just 25 years old, Roth published 'Zur Litteratur und Geschichte des Weda' in 1846 and other writings on the Vedas based mainly on manuscripts (Ibid., p. 449). Interestingly, Roth writes in his book (p. 21) that Buddhism presupposes the Brāhmaṇas and the formation of the ritual, but elsewhere opined that the *Ṛgveda* did not exclusively contain religious hymns (Ibid., pp. 450–452). Böhtlingk and Roth published *Sanskrit-Wörterbuch* in seven parts, 9428 pages total, between 1852 and 1875, for which Roth contributed to the vocabulary of the *Ṛgveda* (Ibid., p. 427). Roth published 'Jāska's Nirukta sammt den Nighaṇṭavas heausegegeben und erlätert von Rudolph Roth' in 1852. Originally in favour of using Indian commentaries, later Roth was against relying on Sāyaṇa's commentary, which interprets through the ritual application and legends; he believed that a text must be understood in and of itself except where the text is not clear (Ibid., pp. 430, 446, 449, 459, 481). In contrast, Theodor Goldstücker argued for Indian interpreters and Müller printed the *Ṛgveda* with Sāyaṇa's commentary (Ibid., pp. 443–444). E. Windisch suggests that it is possible that Roth never sufficiently took into account the sacred character of the hymns (Ibid., p. 460). Adalbert Kuhn studied the cultural connection of the Indogermanic people through languages, myths, and stories (Ibid., pp. 465–467). In connection with Benfey's edition of the *Sāmaveda Saṁhitā* and manuscripts, Weber wrote an essay 'Über die Literatur des Sāmaveda' in 1850. Based on a manuscript, he gave the first precise report on the *Śāṅkhāyana Brāhmaṇa*, drew attention to the legends and sacrifices like the Vājapeya and Rājasūya, and contributed many other important essays. In 1859 Müller published *History of Ancient Sanskrit Literature*, in which chapter 2 discusses the Brāhmaṇa period. He dealt with the Upaniṣads and then the *Śatapatha Brāhmaṇa*, the *Vaṁśa Brāhmaṇa*, ritual passages from the *Aitrareya* and *Kauṣītaki Brāhmaṇa*s, and two famous legends – Śunaḥśepa from the *Aitareya Brāhmaṇa* and the flood from *Śatapatha Brāhmaṇa* (Ibid., pp. 492, 493). In Müller's opinion, the Brāhmaṇa texts misunderstood the original intention of the Vedic hymns.[86] Abel Bergaigne published *Religion Védique* in 1878 and Hermann Oldenberg came out with *Die Religion des Veda* in 1894. A.A. Macdonell wrote *Vedic Mythology* in 1897. Sylvain Lévi published *La doctrine du sacrifice dans les Brāhmaṇas* in 1898.

More progress was made on Brāhmaṇa texts by twentieth century Indologists. To name a few examples, A.B. Keith published a translation of the *Śāṅkhāyana Āraṇyaka* in 1908, an edition and translation of *The Aitareya Āraṇyaka* in 1909, a translation of the *Aitareya* and *Kauṣītaki Brāhmaṇa*s titled *Rigveda Brahmanas* in 1920, and *The Religion and Philosophy of the Veda and Upanishads* in 1925. Raghu Vira published an edition of the first book of the *Jaiminīya Brāhmaṇa* in 1937. His son, Lokesh Chandra, published the rest in 1954. Caland wrote on the Agniṣṭoma in 1906 and published a German translation of 212 long extracts from the *Jaiminīya Brāhmaṇa* in 1919, a Sanskrit edition and English translation of the first seven *kāṇḍa*s of the *Kāṇva Śatapatha Brāhmaṇa* in 1926 (revised by Raghu Vira), and an English translation of the *Pañcaviṃśa Brāhmaṇa* in 1931. Continuing the work of his teacher Caland, the prolific Dutch scholar Jan Gonda wrote extensively on various philological and thematic topics pertaining to Vedic tradition, with special attention to the Brāhmaṇa texts. His books include: *Notes on Brahman* (1950), *The Vision of the Vedic Poets* (1963), *Loka: World and Heaven in the Veda* (1966), *Eye and Gaze in the Veda* (1969), *Vedic Literature* (1975), *The Ritual Sūtras* (1977), *The Mantras of the Agnyupasthāna and the Sautrāmaṇī* (1980), *The Vedic Morning Litany (Prātaranuvāka)* (1981), *The Haviryajñāḥ Somāḥ* (1982), *Prajāpati and the Year* (1984), *Prajāpati's Rise to Higher Rank* (1986), *Mantra Interpretation in the Śatapatha Brāhmaṇa* (1988), *Prajāpati's Relations with Brahman, Bṛhaspati and Brahmā* (1989), and *The Functions and Significance of Gold in the Veda* (1991).[87] With Gonda serving on his PhD thesis committee, G.U. Thite published *Sacrifice in the Brāhmaṇa-Texts* (1975), books on medicine and music in Vedic literature, as well as numerous articles on the Brāhmaṇa texts, rituals, and interpretation. A student of Gonda, Henk Bodewitz translated the first *kāṇḍa* of the *Jaiminīya Brāhmaṇa* (1.1-364) in two volumes, published *The Daily Evening and Morning Offering (Agnihotra)* (1976), and has contributed a great deal to our understanding of the Brāhmaṇa texts. Brian Smith wrote *Reflections on Resemblance, Ritual, and Religion* (1989) and *Classifying the Universe* (1994). Joanna Jurewicz has written extensively about the philosophy of the Brāhmaṇa texts in her book *Fire, Death, and Philosophy* (2016) and in numerous articles. Her groundbreaking work of unpacking the metaphorical language of the Vedas using cognitive linguistics offers a valuable corrective measure to literal interpretation. They, along with Louis Renou, Paul-Emile Dumont, Herman Tull, J.C. Heesterman, Frits Staal, Jan E.M. Houben, Micheal Witzel, and others, stand out among recent scholars on the Brāhmaṇa texts who have provided informative resources to understand middle and late Vedic thought.

Nevertheless, some are still reluctant to pay attention to what these texts say because of the bias inherited from the colonial period. The Enlightenment perspective of that time upheld the primacy of reason, a progressive view of history, universalism, and science. Mythology like that found in the *arthavāda* of the *agnihotrabrāhmaṇa*s was not valued as much as rational thought and, in addition, there was a tendency to interpret Vedic narratives literally, rather than metaphorically as pointing to something beyond the words themselves. Macdonell's *Vedic Mythology* reads very differently from Aurobindo's *The Secret of the*

Vedas.[88] This bias is present even in Vedic pioneers like Müller, who called the Brāhmaṇas 'the twaddle of idiots', (Müller, 1860, p. 389) and Eggeling, who famously wrote in the introduction to his translation of the *Mādhyandina Śatapatha Brāhmaṇa* that aside from a few other works, 'nothing more absurd has probably ever been imagined by rational beings' (Eggeling 1882, p. ix). It is no wonder that scholars of Buddhism often do not understand the relevance of the Brāhmaṇa texts to Buddhist tradition as it developed in India. Others prefer to turn a blind eye and advance their own theories about the doctrine of karma and rebirth originating outside of Vedic tradition. There is resistance to taking the philosophy of the Brāhmaṇa texts, and especially its *arthavāda,* into account because of the rational bias inherited from the Enlightenment, because Mīmāṃsakas rejected the independent validity of the explanatory portions of the Brāhmaṇa texts, which they considered to be only supplementary to injunctions, and because of how the ideas from the *agnihotrabrāhmaṇa*s travelled throughout Indian thought.[89]

Conclusion

The *agnihotrabrāhmaṇa*s advance their own travel theory that starts and ends with the movement from and back into the primordial being. The act of creation itself is a movement from the unconditioned absolute to relative existence in a given *loka* (sphere of experience) – quite literally for sightseeing. Reborn through the Agnyādheya, the sacrificer makes a *loka* and body (*ātmānaṃ sam+√kṛ*) and travels to the *svarga loka* in each performance of the Agnihotra. Yājñavalkya advises how to perform the Agnihotra when abroad in *Kāṇva Śatapatha Brāhmaṇa* 3.1.4.4 and *Jaiminīya Brāhmaṇa* 1.2. And of course, departing the physical body takes the Agnihotrin on another trip, depending on whether he knows in such a way or not. These ideas travelled into the discourses of the Upaniṣads, Buddhist teachings, and the *Bhagavadgītā*, where new concepts and practices were introduced to enliven the tradition. The Śrautasūtras focused on *vidhi* alone, without the *arthavāda,* for didactic purposes, which had unintended consequences. For example, technical ritual practice became a primary focus of later *pauruṣeya* Vedic tradition. Overlooking the basic foundation laid out in the *arthavāda* sections of the *apauruṣeya* Brāhmaṇa texts, both Vedānta and Mīmāṃsā divided the Vedas into two parts, one focused on *karma* and the other on *jñāna*, relegating ritual to a means for the purification of the mind, culminating in knowledge. When the early missionaries and civil servants of the East India Company first encountered the local brahmin pandits, they inadvertently took on interpretations from later Indian tradition without yet having direct access to Vedic texts. This resulted in influential histories, books, and essays being written about Vedic religion before sufficient understanding of the Vedas had been established. These secondary sources, in turn, affected the views of nineteenth century Indologists and intellectuals in general. Significant progress in understanding the Brāhmaṇa texts have been made in the twentieth century, but many scholars still deem this material not worth their while because of conclusions drawn by Indian tradition and orientalists alike. Many prefer to reference Eggeling's translation, the first of the Brāhmaṇa texts translated into English, to find support for their own views on the Brāhmaṇa texts without investing in understanding the genre as a whole. The inattention to the *arthavāda* going back to the time of the Brāhmaṇa texts themselves means that the most obvious continuities in Indian thought are constantly overlooked.

At each turn, Said's (1983, p. 226) travelling theory helps to situate the various contexts to which the philosophy of the *agnihotrabrāhmaṇa*s travelled and the 'processes of representation' that differ 'from those at the point of origin'. This is seen, for example, in the different terms used for 'body' throughout Indian tradition and the tension that ensued between the body and the embodied, even within the same genre. The Brāhmaṇa texts use *ātmán* to refer to the body and considered the Agnihotrin to have two bodies—a physical body and an unmanifest, undying body. The two bodies function like gears to make the world continuous so that the primordial being can experience relative existence for a hundred years. The Upaniṣads continue to use *ātmán* in the sense of body, but begin to see the *ātmán* as the primordial being embodied in the heart. The variegated ways that the Upaniṣads construe *ātmán* and the interpretations that followed are reflected in Buddhist discourses. Buddhism still understands *ātmabhāva* in the sense of a body, but describes all conditioned things as *anātman*. In some teachings, however, the Buddha does not deny an *ātman*. The *Bhagavadgītā* contrasts the body (*deha*) with the embodied (*dehin*) *ātman*. Here, the understanding of *ātman* as body has been replaced with *ātman* as the absolute self. Śaṅkarācārya's Advaita Vedānta emphasizes *ātman* as the absolute self embodied in the body and Mīmāṃsā, similarly, contrasts the *ātman* with the body. Later tradition distinguishes the *jīvātman* from the *paramātman*. Though these expressions refreshed previous ones and the concept of *ātman* was restyled, the basic idea behind the various teachings points to corresponding experiences of reality that may very well be interchangeable. These nuances, however, affect the way a person without direct experience understands how conditioned existence relates to one's true nature.

This travelling doctrine and its developments over time reflect the challenging philosophical task to describe how relative, conditioned existence relates to primordial being – how conditioned knowing relates to unconditioned knowing. Due to the diverse meaning behind and the application of concepts, it is not possible to assume that a term always meant the same thing, even within the same texts. Philosophical concepts have to be periodically enlivened so as to continue to point to something beyond themselves. As the concepts shifted, different understandings of what the concepts point to emerged over time, for example, in Indian tradition when renunciation became a decisive lens for practice and in colonial thought when rational understanding trumped interpreting language metaphorically. The *agnihtorabrāhmaṇa*s take the primordial being as the point of departure: the primordial being becomes man. While later philosophy does not lose sight of the unconditioned, there is a tendency to focus more on trying to escape conditioned existence: man seeks liberation. No longer is conditioned existence seen as a joy ride. Instead, man is seen as trapped in *saṃsāra*.

Through repeated activity that involved knowing in a certain way, Vedic ritual – as envisioned in the *agnihotrabrāhmaṇas* – constructed an embodied knowing, an inclusive knowing that included the two bodies of the Agnihotrin and the three worlds through which the continuity of relative existence is possible. Through ritual practice, the Agnihotrin knew everything (*sarvam*) and ultimately, through knowing in this way, became *sarvam* once again. Despite the *agnihotrabrāhmaṇas'* *arthavāda* being largely neglected because of the *jñāna* vs. *karma* split in the *Bhagavadgītā*, Śaṅkarācārya's Advaita Vedānta, and Mīmāṃsā, many philosophical ideas of the Brāhmaṇa texts persist, sometimes in an almost unrecognizable guise, in later Indian intellectual history and beyond. The Brāhmaṇa texts that have survived recount a time when it was understood that the primordial being itself willingly entered relative existence to experience itself in relation to others and thereby constitutes everything manifest and unmanifest. Though the original mind's passage into relative existence now goes on a little longer than expected, man and everything else around him is no less divine than when the Brāhmaṇa texts explained his origins in connection with the ritual offering (*yajñá*). The very exchange that enables the continuity of relative existence can still be engineered to lead to the highest goal, *sarvam*, for *ya evaṃ veda*.

Notes

1 I am grateful to Mattia Salvini for suggesting that I think about Vedic travel, to Prof. G.U. Thite for answering my endless questions about Vedic tradition, and to Nabanjan Maitra for reading a draft.

2 JB 2.47: *mano vai prajāpatiḥ*; KB 10.1, 26.3: *prajāpatir vai manaḥ*; ŚBM 4.1.1.22: *prajā́patir vai mánaḥ*; TS 2.5.11.5 *mána iva hí prajā́patiḥ*, TS 3.4.7.1: *prajā́patir viśvákarmā mánaḥ*, TS 6.6.10.1: *mána iva hí prajā́patiḥ*; TB 2.2.6.2: *mána iva hí prajā́patiḥ*, 3.7.1.2: *máno vaí prajā́patiḥ*; JUB 1.33; SVB 1.1.1; Gonda, 1983; Lévi, 1898, p. 122; Keith 1914, p. 202.

3 ŚBK 1.2.4.1-6; cf. ŚBM 2.2.4.1-7; Bodewitz, 2003, pp. 16–17.

4 ŚBK 3.1.10.1-4; cf. ŚBM 2.3.3.1-4; Bodewitz, 2003, pp. 18–19.

5 *yájamāno 'gniḥ* | ŚBM 6.3.3.21, 6.4.1.3, 6.4.1.12, 6.4.4.18, 6.5.1.8, 6.6.2.7, 6.7.1.24, and 6.7.3.12. *eṣá vā́ agnír vaiśvānaráḥ* | *yád brāhmaṇáḥ* | TB 2.1.4.29; cf. TB 3.7.3.1-5 and KS 6.6: 56.1ff; Bodewitz, 2003, pp. 137, 116.

6 ŚBK 3.2.5.1-3; cf. ŚBM 11.2.3.1-6.

7 ŚBK 3.2.5.3: *máno vaí rūpaṃ mánasā hí rūpaṃ védedám idáṁ rūpam íti … vā́g vai nā́ma vācā hi nāmābhivyāhárati* |

8 See Smith's chapter 4, which deals with 'The Ritual Construction of Being'.

9 Bodewtiz (1973, p. 222) observes, 'Remarkably this identification with the *prāṇāḥ* hardly occurs in the older passages on the agnihotra (KS.; MS.; TB.)'.

10 JB 1.2: *so 'ta āhutimayo manomayaḥ prāṇamayaḥ…ṛṅmayo yajurmayas sāmamayo brahmamayo hiraṇmayo 'mṛtas sambhavati* |

11 Bodewitz (1973 p. 55, n. 1) cites ŚBM 7.4.2.40 for two wombs and JB 1.259 for two births of the *yajamāna mithunāt* and *yajñāt*.

12 See also ŚBK 3.2.6.2: *āhutimáyam u vā́ etáṁ sukṛtamáyaṃ yájamānasyātmā́naṁ sáṁskurvanti* |

13 For *sam+√kṛ* with *ātman*, see Smith, 1989, pp. 86–92, 101–119; in the context of the Agnicayana, see Jurewicz, 2016, pp. 360–373. Smith cites, for example, AiB 6.27, KB 3.8, ŚBM 11.2.6.13.

14 Gonda (1966, p. 10) observes that the meaning 'to see' developed from 'to give light, to shine', which points to sight relying on light. For example, *locana* means illuminating, the eye.

15 For example, PB 12.11.12 says, *svargalokaḥ puṇyaloko bhavati* and, according to ŚBM 3.6.2.15, 'Therefore they say, "he who has sacrificed is one whose world is merit"' (*tásmād āhuḥ púṇyaloka ījāna íti*). Cf. Eggeling: 'wherefore they say, "He who has sacrificed shares in the world of bliss"'.

16 Or the 'arms of death' (KB 2.9). According to Bodewitz (2019, p. 14), one may assume *punarmṛtyu* was not a Vedic invention, but I see no evidence for this claim.

17 JB 1.5: *dvau haiva samudrāv acaryāv ahaś caiva rātriś ca...tayor vā etayor atyay-anam asti yathā vaiṣavaṃ vā syāt setor vā saṃkramaṇam | astam ite purā tamisrāyai suvyuṣṭāyāṃ purodayāt* | Cf. Bodewitz, 1973, p. 30.

18 For the sun as a wheel, see ṚV 1.164.11, 1.175.4, 4.30.4. For the 'wheel of the sun', see ṚV 4.28.2, 5.29.10. For day and night following each other while changing color, see ṚV 1.113.2.

19 ŚBM 2.3.1.36: 'In the evening he offers Sūrya into Agni, and in the morning he offers Agni into Sūrya' and JB 1.9: 'When the sun sets, it offers itself into the fire...When the sun rises, the fire rises after it. It offers itself in the sun' (Bodewitz, 2003, p. 146).

20 JB 1.8 states:

> Having thus collected it he offers it in the evening. It passes that night in the condition of an embryo. Having collected it in the same way he offers it at dawn. He causes it to be born. Right at its birth it drives away all evil for him. The gods were born from Prajāpati. This one (the sun) is also born from him who knows thus. He wins a 'world' that is as large as the one Prajāpati has one.
>
> (tr. Bodewitz, 1973, p. 36; cf. 2003, pp. 146 and 151–152)

21 Sāyaṇa: *garbharūpeṇāgnāv asthitaṃ sūryaṃ rātriḥ 'tira eva' tirohitam ācchāditam karoti* | 406.

22 The *bahiṣpavamāna* chant is also called a boat (*nauḥ*) bound for *svár* and the virtuous *ṛtvij* priests are the rudders and oars that convey to the far shore (*sampāraṇa*). See ŚBK 5.3.1.8.

23 Cf. ŚBM 2.3.3.15; Bodewitz (2003, p. 58).

24 Bodewitz (2003) provides many examples. In the KB, one reaches the *svarga loka* at twilight and, by touching the coals with the ladle, he places the sacrificer in heaven (44, 110). In GB 1.3.12, by the first libation, the sacrificer places himself in heaven and by looking back at the *gārhapatya* fire, he keeps in contact with this world (103). In ŚBM 2.5.3.6-7, looking back toward the *gārhapatya* joins this and yonder world, whereas by the second libation the sacrificer places himself in heaven (103).

25 According to AiB 7.10, the sacrificer ascends to the *svarga loka*.

26 The texts further describe one reaching, ascending, securing, discerning (*pra+√jñā* TS 6.5.8.2), approaching, turning oneself to, etc. *svarga* (Gonda, 1966, p. 91 n. 11). Cf. TB 2.6.10.4: *aspṛkṣat...dívam* | TB 3.6.13.1: *dyāṃ ... aspṛkṣat* |

27 ŚBM 3.3.4.3: *tád enemā́m lokā́n ā́spṛṇoti tásya hi ná hantā́sti ná badho yénemé lokā́ ā́spṛtāḥ* | Gonda (1966, p. 38) explains that by reciting in a rite VS 4.30, one gains these worlds 'for there is no slayer, no deadly shaft for him by whom 'these worlds' have been gained...'.

28 TB 1.2.1.2: *śatáṃ jīvema śarádaḥ sū́vīrāḥ...śatúṃ jīvema śarúdaḥ purúcīḥ* |

29 See also ŚBK 3.1.7.1.

30 Bodewitz (1973, p. 65 n. 11) already observed that the ŚBM 11.3.1.8 has *viśvā jatāni yo 'bibhaḥ* instead of JB's *sa vidvān pravasan vide*.

31 ŚBK 1.4.2.3 and 7: *gṛhā vai gā́rhapatyaḥ*. *Gṛha* is used in the plural, even though there is only one house, from which the other fires are taken. *Gārhapatya* is an abstract form of *gṛhapáti* ('lord of the house'), which in the *Ṛgveda* is used only to describe Agni (Jamison, 2019, p. 79).

32   See also Jamison and Brereton, 2014, p. 783 and Bausch, 2020.

33   Sāyaṇa, concerned with how to keep up the obligation to perform the Agnihotra twice a day for life, asks how the fault of staying abroad can be avoided (*asya pravasato yajamānasya anapaproṣitam pravāsadoṣābhāvaḥ*).

34   PB 10.4.4 also says that the breath is always awake. See also Bodewitz, 1973, p. 242, n. 12.

35   ŚBK 3.1.11.6: *yájamāna evá prāṇo yā́vad dhy eva yájamānaḥ prāṇéna prā́ṇiti tā́vad evá juhoty átha yadā́ prā́ṇo 'pakrā́maty átha vyávacchidyate |*

36   ŚBK 3.1.12.28: *sa yád etām ā́hutiṁ satī́ṁ nā́gnau júhvaty éṣv evá prāṇéṣu hūyate |* See also ŚBM 11.5.3.8-12 in which the brahmin offers into himself, into his *prāṇāḥ*, which substitute for the fires (Bodewitz, 2003, pp. 137–138) and KB 2.8 in which there is a permanent offering in each other of the fire and sun, night and day, in- and exhalation. He who knows these six sacrifices performs the Agnihotra, even if he does not offer in the fire (*sa ya etāni ṣaḍ juhvati veda, ajuhvata evāsyāgnihotraṁ hutaṁ bhavati*), though Bodewitz asserts that the actual performance of the rite is not wholly abolished (Bodewitz, 1973, p. 240).

37   See also Smith, 1989, pp. 112–117.

38   Cf. VādhS 3.4: 'The son becomes a father. It has been said (in ṚV 1.89.9) "O gods, there are hundred autumns…" Just as here in this world man behaves towards the fires, even so the fires behave towards him in yonder world' (tr. Bodewitz, 2003, p. 171).

39   See also 23 n. 20 and ŚBM 12.5.2.13 on the funeral ritual.

40   Translation emended from: 'He (arises) from this (fire) and…' Cf. ŚBK 3.1.9.3, 3.2.6.2; ŚBM 11.2.6.13: *sá ṛ̀nmáyo yajurmáyaḥ sāmamáya āhutimáyaḥ svargáṁl lokám abhisámbhavati.*

41   *tád dhedaṃ tarhy ávyākṛtam āsīt | tán nāmarūpā́bhyām eva vyā́kriyatāsau nā́māyám idáṃrūpa íti | tád idam ápy etárhi nāmarūpā́bhyām eva vyā́kriyata asau nā́māyam idáṃrūpa íti | sá eṣá iha právistaḥ ā́ nakhāgrébhyo yáthā kṣuraḥ kṣuradhāné 'vahitaḥ syād viśvambharó vā viśvambharakulāye | taṃ na páśyanty ákṛtsno hi sá prāṇánn evá prāṇo nā́ma bhávati | vádan vā́k páśyaṃś cákṣuḥ śṛṇvañ chrótram manvāno mánas tā́ny asyaitā́ni kármanāmāny eva sa yó 'ta ékaikam upā́ste na sá vedā́kṛtsno hy eṣó 'ta ékaikena bhávaty ātméty evópāsītā́trā hy ete sárva ékaṃ bhávanti | tád etát padanī́yam asya sárvasya yád ayám ātmānénā hy etat sárvaṃ véda yáthā ha vaí padénānuvindéd eváṃ kīrtiṃ ślókam vindate yá evaṃ véda || BĀUK 1.4.7 ||*

42   BĀUK 4.3.7: *katamá ātméti – yo 'yáṃ vijñānamáyaḥ prāṇéṣu hṛ́dy antárjyotiḥ pū́ruṣaḥ'.* 4.4.22: *sa vā́ eṣá mahān ajá ātmā́ yo 'yáṃ vijñānamáyaḥ prāṇéṣu yá éṣo 'ntarhṛ́daye.* Cf. ChU 8.3.3 and KU 2.20.

43   See TS 3.10.8.9, ŚĀ 11.6: *ātmā hṛdaye |*

44   See also *ásmāc chārīrā́d ātmánaḥ* at BĀUK 4.2.3.

45   Cohen (2018, p. 288) explains that in *Taittirīya Upaniṣad* 1.6.1, *puruṣa* is used for the immortal being, made of *manas* and dwelling in the heart.

46   See Cohen, 2018, pp. 328–329.

47   Cohen (2018, pp. 343–344) mentions that the vulgate *Maitrī Upaniṣad* consists of six chapters, whereas the Southern recension has only four. While *ātman* is central in the Southern recension, it is hardly mentioned outside of those passages.

48   ChUBh 6.16.3: *tadvat tasyāpi dehādiṣvātmabuddhitvān na syāt sadātmavijñānam | tasmād vikārānṛtādhikṛtajīvātmavijñānanivartakam evedam vākyaṃ tat tvam asīti sid-dhamiti |* Cf. Gambhīrānanda, 2003, p. 502.

49   Sāyaṇa on ṚV 1.164.20: *atra laukikapakṣidvayadṛṣṭāntena jīvaparamātmānau stūyete…vṛkṣo dehaḥ…tayoḥ anyaḥ jīvātmā pippalaṃ karmaphalam svādubhūtam atti bhuṅkte | yasya yad upārjitaṃ tat tasya svādu bhavati | anyaḥ paramātmā anaśnan āpnakāmatvena abhuñjānaḥ spṛhāyāḥ abhāvāt |*

50   Prāṇa is the undying (BĀUK 1.6.3: *prāṇo vā́ amṛtam*). Cf. *Maitrī Upaniṣad* 6.9: *prāṇo 'gniḥ paramātmā vāi pañcavāyuḥ samāśritaḥ, sa prītaḥ prīṇātu viśvam viśvabhuk |*

51   The *ātman* is established in the breath at BĀUK 3.9.25.

52 BĀUK 4.3.7: *katamá ātméti – yo yáṃ vijñānamáyaḥ prāṇéṣu hṛ́dy antárjyotiḥ púruṣaḥ.*
    4.4.22: sa vā́ eṣá mahān ajá ātmā́ yo 'yáṃ vijñānamáyaḥ prāṇéṣu yá éṣo 'ntarhṛ́daya
    ākāśas tásmiñ chete | sárvasya váśī | sárvasyéśānaḥ sárvasyādhipatiḥ sa ná sādhunā
    kármaṇā bhū́yán no evāsádhunā kánīyán |

53 See also the talk by Joanna Jurewicz at the Greater Magadha Symposium hosted by the
    University of Alberta in May 2021 in which she demonstrates how Yājñavalkya's ideas
    are prefigured in the *Śatapatha Brāhmaṇa.* Jurewicz shows how Yājñavalkya's teaching
    on speech in BĀU 4.1ff is based on ŚBM 10.5.1.1ff. In this talk, she cogently shows the
    continuity of ŚBM 10.5.1.3 with BĀU 1.2.3 and ŚBM 6.1.1.2 with BĀU 4.2.2.

54 See, for example, Tull's discussion on the fate of the Pārikṣitas in BĀU 3.3.1-2, which
    has antecedents in ŚBM 13.5.4.1-3.

55 For more details on historical developments in *śrauta* ritual practice and interpreta-
    tion—including the Vedāṅgas, darśanas, and Smārta texts—see Bausch, forthcoming.

56 G.U. Thite, retired professor from the University of Poona, discussion in 2021.

57 Sn 249: *Na macchamaṃsaṃ nānāsakattaṃ na naggiyaṃ muṇḍiyaṃ jaṭā jallaṃ*
    kharājināni vā nāggihuttass'upasevanā va yā ye vā pi loke amarā bahū tapā mantāhutī
    yañña-m-utūpasevanā sodhenti maccaṃ avitiṇṇakaṃkhaṃ.

58 In this way, the sacrificers (*yājaka*) fell from righteousness (*dhamma,* Sn 312).

59 Sn 506: *Yajassu, yajamāno Māghā ti Bhagavā sabbattha vippasādehi cittaṃ:*
    ārammaṇaṃ yajamānassa yaññaṃ, ettha patiṭṭhāya jahāti dosaṃ.

60 In Buddhism, *saṃsāra* is qualified as a fourfold flood (*kāma-, bhava-, diṭṭhi-,* and *avi-*
    jja-ogha).

61 Jurewicz (2000) describes the connection between *pratītyasamutpāda* and Vedic
    thought more broadly, whereas I restrict myself to the Brāhmaṇas and, for the most
    part, to the *agnihotrabrāhmaṇa*s.

62 Sn 168: '*Kismiṃ loko samuppanno, iti Hemavato yakkho*
    kismiṃ kubbati santhavaṃ, kissa loko upādāya kismiṃ loko vihaññati'.
    Sn 169: '*Chassu loko samuppanno, Hemavatā ti Bhagavā*
    chassu kubbati santhavaṃ, channam eva upādāya chassu loko vihaññati'.
    SN 1.7.10 (PTS p. 41): *Kismiṃ loko samuppanno || kismiṃ kubbati santhavaṃ ||*
    kissā loko upādāya || kismiṃ loko vihaññatīti ||
    Chasu loko samuppanno || chasu kubbati santhavaṃ ||
    channam eva upādāya || chasu loko vihaññatīti ||

63 I am grateful to Bhikkhu Bodhi for bringing this passage to my attention and for teach-
    ing MN 22 in Pāli class.

64 MN 22: *Siyā nu kho bhante ajjhattaṃ asati paritassanā ti. – Siyā bhikkhūti Bhagavā*
    avoca. Idha bhikkhu ekaccassa evaṃ diṭṭhi hoti: so loko so attā, so pecca bhavissāmi
    nicco dhuvo sassato avipariṇāmadhammo, sassatisamaṃ tath' eva ṭhassāmīti. So
    suṇāti Tathāgatassa vā Tathāgatasāvakassa vā sabbesaṃ diṭṭhiṭṭhānādhiṭṭhāna-
    pari-yuṭṭhānābhinivesānusayānaṃ   samugghātāya   sabbasaṅkhāra-samathāya
    sabbūpadhipaṭinissaggāya taṇhakkhayāya virāgāyanirodhāya nibbānāya dhammaṃ
    desentassa. Tassa evaṃ hoti: Ucchijjissāmi nāma su, vinassissāmi nāma su, na su nāma
    bhavissāmīti. So socati kilamati paridevati, urattāḷiṃ kandati, sammohaṃ āpajjati.

65 BhG 13.30: *yadā bhūtapṛthagbhāvam ekastham anupaśyati | tata eva ca vistāraṃ*
    brahma saṃpadyate tadā |

66 BhG 8.3: *akṣaraṃ brahma paramaṃ;* 8.21 *avyakto kṣara iti.*

67 BhG 9.16: *ahaṃ kratur ahaṃ yajñaḥ svadhāham aham auṣadham | mantro ham aham*
    evājyam aham agnir ahaṃ hutam |

68 BhG 14.3: *mama yonir mahad brahma tasmin garbhaṃ dadhāmy aham | saṃbhavaḥ*
    sarvabhūtānāṃ tato bhavati bhārata |

69 BhG 4.26: *śrotrādīnīndriyāṇy anye saṃyamāgniṣu juhvati | śabdādīn viṣayān anya*
    indriyāgniṣu juhvati |

70 BhG 4.27: *arvāṇīndriyakarmāṇi prāṇakarmāṇi cāpare | ātmasaṃyamayogāgnau juh-*
    vati jñānadīpite |

71 BhG 4.28: *dravyayajñās tapoyajñā yogayajñās tathāpare | svādhyāyajñānayajñāś ca yatayaḥ saṃśitavratāḥ |*

72 BhG 4.30: *apare niyatāhārāḥ prāṇān prāṇeṣu juhvati | sarve 'py ete yajñavido yajñakṣapitakalmaṣāḥ |*

73 BhG 2.46: *yāvān artha udapāne sarvataḥ samplutodake | tāvān sarveṣu vedeṣu brāhmaṇasya vijānataḥ |*

74 See Śaṅkarācārya on BhG 2.10 (*agnihotrādiśrautasmārtakarmasahitāt jñānāt ... agnihotrādikarma*), 6.10 (*agnihotrādikaṃ karma karoti*), and 11.48 (*kriyābhir agnihotrādibhiḥ śrautādibhiḥ*).

75 Śaṅkara on BhG 2.10: *yathā ca svargādikāmārthino 'gnihotrādikarmalakṣaṇadharmānu ṣṭhānāya āhitāgneḥ kāmye eva agnihotrādau pravṛttasya sāmi kṛte vinaṣṭā 'pi kāme tad eva agnihotrādyanutiṣṭhato 'pi na tatkāmyam agnihotrādi bhavati | tathā ca darśayati bhagavān – kurvann api na lipyate na karoti na lipyate iti* [see 5.7] *tatra tatra ||*
    yac ca pūrvaiḥ pūrvataraṃ kṛtam karmaṇaiva hi saṃsiddhim āsthitā janakādayaḥ iti, tat tu pravibhajya vijñeyam | tat katham ? yadi tāvat pūrve janakādayaḥ tattvavido 'pi pravṛttakarmāṇaḥ syuḥ, te lokasaṃgrahārtham guṇā guṇeṣu vartante iti jñānenaiva saṃsiddhim āsthitāḥ, karmasaṃnyāse prāpte 'pi karmaṇā sahaiva saṃsiddhim āsthitāḥ, na karmasaṃnyāsaṃ kṛtavanta ity arthaḥ | atha na te tattvavidaḥ | īśvarasamarpitena karmaṇā sādhanabhūtena saṃsiddhiṃ sattvaśuddhim, jñānotpattilakṣaṇāṃ vā saṃsiddhim, āsthitā janakādaya iti vyākhyeyam |

76 Śaṅkarācārya on BhG 2.10: *tasmād gītāśāstre kevalād eva tattvajñānān mokṣaprāptiḥ na karmasamuccitāt iti niścito 'rthaḥ |*

77 Śaṅkarācārya's introduction to chapter 3: *jñānakarmaṇoḥ samuccayānupapattiḥ | tasmāt kevalād eva jñānān mokṣa ity eṣo 'rtho niścito gītāsu sarvopaniṣatsu ca |*

78 Śaṅkarācārya on BhG 3.3: *tasmāt kayāpi yuktyā na samuccayo jñānakarmaṇoḥ |*

79 *tasyāś ca jñānaniṣṭhāyāḥ saṃnyāsinām evānuṣṭheyatvam, bhinnapuruṣānuṣṭheyatvavacanāt |*

80 Śaṅkarācārya on BhG 3.4: *na karmaṇā kriyāṇāṃ yajñādīnām iha janmani janmāntare vā anuṣṭhitānām upāttaduritakṣayahetutvena sattvaśuddhikāraṇānāṃ tatkāraṇatvena ca jñānotpattidvāreṇa jñānaniṣṭhāhetūnām, jñānam utpadyate puṃsāṃ kṣayāt pāpasya karmaṇaḥ |*

81 Śaṅkarācārya on BhG 3.4: *karmārambhasyaiva naiṣkarmyopāyatvāt | na hy upāyam antareṇa upeyaprāptir asti | karmayogopāyatvaṃ ca naiṣkarmyalakṣaṇasya jñānayogasya, śrutau iha ca pratipādanāt |* I am grateful to G.U. Thite for discussing this point with me on 10 June 2021.

82 Alexander Dow also read Hinduism favorably as monotheism, but attacked Holwell's account. See Trautmann, 1997, pp. 70, 72.

83 (Windisch, 2008: 49).

84 The last three volumes of Müller's edition are dedicated to Queen Victoria. For volume 2, Müller received support from Ballantyne, the Principal of the Sanskrit College at Benares, and from F. Edavard Hall in Benares. After the 500 copies of the first edition sold out, the second edition was printed under the patronage of the Mahārāja of Vijayanagara (Windisch, 2008, pp. 480–482).

85 Windisch (2008, p. 289). vol. 2 1853, vol. 3 1858, vol. 4 1861, 2nd edn vol. 1 1867, and 2nd edn vol. 2 1873.

86 Windisch (2008, p. 494). In 1899, the year before he died, Müller published a book on *The Six Systems of Indian Philosophy.*

87 Gonda translated the *Ṛgvidhāna* in 1951. His articles and books are replete with translations of Vedic passages.

88 The writings and translations in the 1956 volume were originally published in the monthly review *Arya* between 1914 and 1920.

89 I am grateful to Elisa Freschi for clarifying the point about Mīmāṃsā, which is explained in more detail in my forthcoming book.

References

Bausch, L.M. (2020). 'Philosophy of Language in the R̥gveda,' Annals of the Bhandarkar Oriental Research Institute XCVII. Ed. S.S. Bahulkar and Shilpa Sumant: 24–56.

Bausch. L.M. (forthcoming). *The Brāhmaṇa Texts as Philosophy.*

Bhagavadgītā with the Commentary Ascribed to Śaṁkara (Adhyāyas 1–17) etext. Input by Gaudiya Grantha Madira 7/31/2020 on GRETIL.

Bhattacharya, K. (2015). *The Atman-Brahman in Ancient Buddhism.* Canon Publications. Originally: *L'Ātman-Brahman dans le Bouddhisme ancien*, Publications de l'École française d'Extrême-Orien, vol. 90, 1973.

Bodewitz, H. W. (1976). *The Daily Evening and Morning Offering (Agnihotra) According to the Brāhmaṇas.* Brill. Reprinted Motilal Banarsidass, 2003.

Bodewitz, H. W. (Trans.). (1973). *Jaiminīya Brāhmaṇa I, 1–65: Translation and Commentary with a Study Agnihotra and Prāṇāgnihotra.* Brill.

Bodewitz, H. W. (2019). *Vedic Cosmology and Ethics: Selected Studies.* Brill.

Caland, W. (Trans.). (1931). *Pañcaviṃśa-Brāhmaṇa: The Brāhmaṇa of Twenty Five Chapters.* The Asiatic Society. Republished in 1982.

Chāndogya Upaniṣad: With the Commentary of Śaṅkarācārya. (2003). Tr. Swāmī Gambhīrānanda. Advaita Ashrama.

Chandra, L., & Vira, R. (Eds.). (1986). *Jaiminīya Brāhmaṇa of the Sāmaveda*, 2. Motilal Banarsidass Publishers.

Cohen, S. (Ed.). (2018). *The Upaniṣads: A Complete Guide.* Routledge.

Colebrooke, H. T. (1873). 'A Discourse Read at a Meeting of the Royal Asiatic Society of Great Britain and Ireland, on the 15th of March, 1823.' In E. B. Cowell (Ed.), *Miscellaneous Essays* (p. 2). Trübner & Co.

Deleuze, G., & Guattari, F. (1994). *What is Philosophy?* (H. Tomlinson & G. Burchell, Trans.). Columbia University Press.

Dodson, M. (2007). *Orientalism, Empire, and National Culture: India, 1770–1880.* Palgrave Macmillan.

Eggeling, J. (Trans.). (1882–1900), repr. 1972. *The Śatapatha-brāhmaṇa, According to the Text of the Mādhyandina School.* Sacred Books of the East, 12, 26, 41, 43, 44. Motilal Banarsidass Publishers.

Eggeling, J. (Trans.). (2008). *The Śatapatha-Brāhmaṇa: Sanskrit Text with English Translation and Notes* (M. Deshpande, Ed.). New Bharatiya Book Corporation.

Feer, M. L. (Ed.). (1884). *Saṃyutta-Nikāya of the Sutta-piṭaka* (Part 1). Pali Text Society. Repr. 1991.

Freiberger, O. (1998). 'The Ideal Sacrifice. Patterns of Reinterpreting Brahmin Sacrifice in Early Buddhism.' *Bulletin D'Études Indiennes, 16*, 39–49.

Gambhīrānanda. (Trans.). (2003). *Bhagavadgītā: With the Commentary of Śaṅkarācārya.* Advaita Ashrama.

Gonda, J. (1955). 'Reflections on *sarva-* in Vedic Texts.' *Indian Linguistics*, 16 (Chatterji Jubilee Volume), 53–71. Repr. *J. Gonda Select Studies*, 2, 497–515. E.J. Brill, 1975.

Gonda, J. (1966). *Loka: World and Heaven in the Veda.* N.V. Noord-Hollandsche Uitgevers Maatschappij.

Gonda, J. (1969/1975). 'Āyatana,' Adyar Library Bulletin 23: 1-79. Repr. J. Gonda Select Studies, Vol. 2: 178–256. E.J. Brill.

Gonda, J. (1977). *The Ritual Sūtras*, 1, Fasc. 2. Otto Harrassowitz.

Gonda, J. (1983). 'The Creator and His Spirit: Manas and Prajāpati,' Wiener Zeitschrift für die Kunde Südasiens 27: 5–42.

Hegel, G. W. F. (1956). *The Philosophy of History* (J. Sibree Trans.). Dover Publications, Inc.

Hegel, G. W. F. (1975). *Aesthetics: Lectures on Fine Art* (T. M. Knox, Trans., Vol. I). Oxford University Press.

Hegel, G. W. F. (1977). *Phenomenology of Spirit* (A. V. Miller, Trans.). Oxford University Press.

Hirst, J. S. (2018). *Ātman* and *Brahman* in the Principal Upaniṣads. In S. Cohen (Ed.), *The Upaniṣads: A Complete Guide* (pp. 107–120). Routledge.

Jamison, S. W. (2019). 'Vedic ritual: The Sacralization of the Mundane and the Domestication of the Sacred.' In L. M. Bausch (Ed.), *Self, Sacrifice, and Cosmos: Vedic Thought, Ritual, and Philosophy* (pp. 41–55). Primus Books.

Jamison, S. W., & Brereton, J. P. (Trans.). (2014). *The Rigveda: The Earliest Religious Poetry of India* (pp. 1–3). Oxford University Press.

Jha, G. (Trans.). (1933). *Shabara-bhāṣya.* Oriental Institute.

Jones, C. V. (2021). *The Buddhist Self: On Tathāgathagarbha and Ātman.* University of Hawai'i Press.

Jurewicz, J. (2000). 'Playing with Fire: The Pratītyasamutpāda from the Perspective of Vedic Thought.' *Journal of the Pali Text Society*, 26, 77–103. Repr. *Buddhism: Critical Concepts in Religious Studies*, I, P. Williams, 2005.

Jurewicz, J. (2016). *Fire, Death and Philosophy: A History of Ancient Indian Thinking.* Dom Wydawniczy Elipsa.

Jurewicz, J. (2019). 'The Consistency of Vedic Argument.' In L. M. Bausch (Ed.), *Self, Sacrifice, and Cosmos: Vedic Thought, Ritual, and Philosophy* (pp. 41–55). Primus Books.

Keith, A. B. (Trans.). (1914). *The Veda of the Black Yajus School Entitled Taittirīya Sanhitā, part 2.* Harvard University Press.

Keith, A. B. Trans. (1920). *The Rigveda Brahmaṇas: The Aitareya and Kauṣītaki Brāhmaṇas of the Rigveda.* Harvard University Press.

Krick, H. (1982). *Das Ritual der Feuergründung (Agnyādheya).* Österreichische Akademie der Wissenschaften.

Lévi, S. 1898. *La doctrine du sacrifice dans les Brāhmaṇas.* Ernest Leroux.

Majjhima-Nikāya. (1888). 1(V). *Trenckner.* Pali Text Society.

Müller, F. M. (1860). *A History Of Ancient Sanskrit Literature: So far as it illustrates the primitive religion of the Brahmans* (2nd ed.). Williams and Norgate.

Navathe, P.D. (1980). *Agnihotra of the Kaṭha Śākhā: Kāṭhaka Saṃhitā 6.1–9; 7.1–11.* University of Poona.

Pimplapure, G. W. (Ed.). (2002). *Kāṇva Śatapatha: A Critical Edition* (2nd ed.). Maharṣi Sāndīpani Rāṣṭriya Vedavidyā Pratiṣṭhānam.

Potdar, K. R. (1953). *Sacrifice in the Rig Veda*, Bombay: Bharatiya Vidya Bhavan.

Renou, L. (1952). 'On the Word *ātman.*' *Vāk*, 2, 151–157.

Said, E. (1978). *Orientalism.* Pantheon Books.

Said, E. (1983). *The World, the Text, and the Critic.* Harvard University Press.

Schwab, R. (1984). *The Oriental Renaissance: Europe's Rediscovery of India and the East 1680–1880* (G. Patterson-Black & V. Reinking, Trans.). Foreword by Edward Said. Columbia University Press.

Sharma, C. (1987). *A Critical Survey of Indian Philosophy.* Motilal Banarsidass.

Smith, B. (1989). *Reflections on Resemblance, Ritual, and Religion.* Oxford University Press. Repr. Motilal Banarsidass Publishers, 1998.

The Śatapatha-brāhmaṇa, According to the Mādhyandina Recension with the Commentary of Sāyaṇa and Hariśvāmin. (2002). Rashtriya Sanskrit Sansthan.

The Śūraṅgama Sūtra: With Excerpts from the Commentary of the Venerable Master Hsüan Hua. (2009). Buddhist Text Translation Society.

Thite, G. U. (1975). *Sacrifice in the Brāhmaṇa-texts.* University of Poona.

Thite, G.U. (Ed.). (2012). *Kṛṣṇayajurvedīyam Taittirīyabrāhmaṇam: Śrīmatsāyaṇācāryavir acitabhāṣyasametam.* Two volumes. New Bharatiya Book Corporation.

Trautmann, T. (1997). *Aryans and British India.* University of California Press.

Tull, H. W. (1989). *The Vedic Origins of Karma: Cosmos as Man in Ancient Indian Myth and Ritual.* State University of New York Press.

Verpoorten, J. (1987). *Mīmāṃsā Literature. A History of Indian Literature* 6 (fasc. 5). Otto Harrassowitz. Republished Manohar, 2020.

Weber, A. (Ed.). (1855). *The Çatapatha-Brâhmaṇa in the Mâdhyandina-çâkhâ with Extracts from the Commentaries of Sāyaṇa, Harisvâmin and Dvivedaganga.* Williams and Norgate.

Wilson, H. H. (1840–1848). *History of British India: From 1805–1835.* James Madden and Company.

Windisch, E. (2008). *History of Sanskrit Philology and the Indian Archaeosophy* (G. U. Thite, Trans., Vol. 1–2). New Bharatīya Book Corporation.

List of Abbreviations

AiĀ: Aitareya Āraṇyaka | *AiB*: Aitareya Brāhmaṇa | *AV*: Atharvaveda | *BĀUK*: Kāṇva Bṛhadāraṇyaka Upaniṣad | *BhG*: Bhagavadgītā | *ChU*: Chāndogya Upaniṣad | *ChUBh*: Chāndogya Upaniṣad Bhāṣya | *GB*: Gopatha Brāhmaṇa | *JB*: Jaiminīya Brāhmaṇa | *JUB*: Jaiminīya Upaniṣad Brāhmaṇa | *KB*: Kauṣītaki Brāhmaṇa | *KS*: Kāṭhaka Saṃhitā | *KU*: Kāṭha Upaniṣad | *MN*: Majjhimanikāya | *MS*: Maitrāyaṇī Saṃhitā | *MuU*: Muṇḍaka Upaniṣad | *JMS*: Jaimini's Mīmāṃsā Sūtra | *PB*: Pañcaviṃśa Brāhmaṇa | *PTS*: Pali Text Society | *ṚV*: Ṛgveda | *ŚBK*: Kāṇva Śatapatha Brāhmaṇa | *ŚBM*: Mādhyandina Śatapatha Brāhmaṇa | *ŚBh*: Śabarabhāṣya | *Sn*: Suttanipāta | *SN*: Saṃyuttanikāya | *TĀ*: Taittirīya Āraṇyaka | *TB*: Taittirīya Brāhmaṇa | *TS*: Taittirīya Saṃhitā | *TU*: Taittirīya Upaniṣad | *VādhS*: Vādhūla Sūtra

The State, Polity and the Religious

4 Travel(ing) to Write

Authorship and Agency in an Era of Inter-polity Mobility

Rafia Khan

Historiography on medieval travel and mobility has revealed the presence of myriad indigenous literary-performative cultures (eighth-eighteenth centuries) resting on the figure of the semi-itinerant author-performer and/or mythologist like the *bhats* of medieval Rajputana. The 'advent' of the Islamicate in the Indian subcontinent with its literal and imaginative fashioning of the religious-political acts of *hajj* (pilgrimage), *hijra* (emigration), *ziyara* (visit to shrines) on the one hand, and *rihla* (travel for learning and other purposes) on the other hand, also provided a plethora of models for the imagination and act of travel, either inspired religiously, informed by cultural-material incentives or a combination of socio-political roles and statuses. There is thus no gain saying that travel and mobility transcended medieval sub-continental societies and polities.

And yet, this image is crowded with the characters of the itinerant Sufi, the warrior godmen or *ghazi*, the itinerant author-poet-performer, and/or the mercenary, soldiering class. The preponderance of these avid and renowned 'free' travellers in the broader historiographical conversation on travel and mobility in medieval India also informs our conceptual understanding of medieval mobilities, which consequently ends up being characterized as primarily religious-cultic, nomadic, and often, outside the purview or independent of the state/s and its paraphernalia. Historical figures who shape our understanding of medieval travel and movement of a continuous nature also often occupy the micro-meso levels of the medieval body politic or are outside of it. They also typically represent traits of local, sub-local, vernacular, and folkish traditions.

My intention in this chapter is to broaden our understanding of the nature of mobilities in medieval India by tracing the travel and travails of two members of the medieval literati (Shihab Ḥākim and *kavi* Māhēśa), court chroniclers who wrote in the 'classical' languages of the 'high' culture of the medieval courts of Malwa and Mewar, respectively; they sought and closely followed state patronage and the trail of its political penetration while being representatives *par excellence* of the state-dependent elite or ruling class. As the antithesis of the 'modern' visualization of medieval travel and mobility, this socio-political class, unsurprisingly, has been rarely studied as 'travellers'. Since men like Ḥākim and Māhēśa don't typically constitute modern discussions on medieval mobilities, there is obviously no discussion on how these state-dependent members of the elite, especially as they

DOI: 10.4324/9781003544524-7

were court chroniclers, grappled with this constancy of travel and movement and the implications it may have had on their lives and works.

Albeit works on authors and authorship abound in the historical works on medieval Indian sultanates and travel writers in medieval India have also started to receive some much-needed attention, travel(ing) writers have not really been identified as a distinct socio-historical category for study. One reason for this is that almost 'all' authors, as part of a world of mobility and movement, had experienced travel and movement at some time, whether it was from one court to another or from one region to another, thus negating the need for a distinct analytical lens for a group subsumed within various other ruling elite groups. Travel(ing) writers also escape specialized scrutiny from scholars studying travel and mobility because the nature of these journeys associated with the ruling elite is generally seen as periodical, transitory, and not as a way of life or lifestyle which would trigger changes in an individual's work, world, or worldview. This work, however, is different in that it endeavours to show that these movements were neither temporary nor periodical even if the authors themselves may have wanted us to believe so. It also recognizes the authors' journeys through changing contexts and their awareness of the to-and-fro patterns of their lives as having contributed to the form and shape of their work to some extent. Finally, in keeping with the core theme of this volume, these journeys and the distinct experiences of these journeys and the responses generated thereof will deepen our understanding of the nature of travel and mobility in medieval India.

Writers, Travel-writers, and Travel(ing) Writers in Medieval India

At different times in the first half of the thirteenth century, Muhammad Awfi (fl. 1204–1232/1233), a fine orator and a prolific writer, Sadr al-Din Hasan Nizami (fl. 1205), the author of the Taj-ul-Maasi'r, and Qazi Minhaj-us-Siraj Juzjani (fl. 1193–1260), a religious scholar-author and a member of the Ghurid *ulema*, left their homes to move to the nether parts of the Ghurid kingdom in the northern part of the Indian subcontinent. This was a movement away from the centre of Islam, the *dar-ul-Islam*, the 'civilised' and the 'prosperous' home of Islam. These were necessitated departures, prompted by political exigencies, away from the land, the people, and the world they had known. In short, these were separations from their 'defining context of recognitions' (Leed, 1991, p. 34).

Similar sojourns were undertaken by members of the Persian literati at the end of the fourteenth and early fifteenth centuries when the political edifice of the Sultanate in Delhi crumbled. Members of the Persian literati (especially court chroniclers), in the crosshairs of the political turmoil, endured the phenomenon of 'necessitated departure' yet again. Muhammad Bihamad Khani's father travelled from Hazrat-i-Dihli to the nascent capital of Kalpi (Jalaun, Uttar Pradesh), far to the east of the erstwhile glory of the Delhi Sultanate. The whereabouts of Shihab Hakim (b. 781/1379–1380), a litterateur who wrote for the Sultan of Jaunpur (Jaunpur, Uttar Pradesh) but was travelling to the capital of the Malwa Sultanate in Mandu (Dhar, Madhya Pradesh) in the second half of the fifteenth century, are even more difficult to ascertain.

Travel-writers, different from the travel(ing) writers above, also constituted a category of medieval travellers, but one must distinguish between the two largely distinct groups and their activities in medieval India. Travel-writers considered and often labelled their work with 'travel' nomenclature such as *safar* or *rihla*, thus justifying their characterization as travel-writers and their work as travel-texts. Moreover, these authors were considerably conscious of their identities as travellers and their act of travel as their primary social action; for example, Ibn Battuta, who travelled through the Delhi Sultanate in the times of Sultan Muhammad-bin-Tughluq (1325–1351), informs that 'strangers' and 'foreigners' were shown affection and singled out by the sultan with the epithet *azīz* or honourable attached to their name; the organizing principle of a travel-writer's text reveals a dominant 'travel' idiom or theme; for instance, while a travel-writer weaves an account of his own journey in a travel-text, thus allowing the historian to study the travel-writer directly on his journey at various junctures, information on the life and travels of a travel(ing) author can only be detected indirectly through the prism of asides found in prefaces, biographical notes, or sketchy references within larger narratives; and finally, as pointed out by Alam and Subrahmanyam, travel-texts typically carried 'an explicit, even philosophical, reflection on the meaning of travel' (Alam & Subrahmanyam, 2009, p. 21). While it may seem apposite to include such historical material in a work on de-colonizing travel and vernacular mobilities, the awareness travel-writers have of travel as their primary social action and the singularity with which they (are deemed to have) pursued the social action of 'travel', tends to compartmentalise travel as an activity distinct from the usual goings-on of medieval life. This exceptional, almost specialized, practice of 'travel' tends to present a visual of medieval travel as a separate, even isolated phenomenon. This perspective on travel, however, is challenged by that quotidian, everyday form of travel which one finds in the lives of travel(ing) authors and which indicates that mobility was embedded in a motley of socio-political, economic, and religious-cultural contexts in the medieval subcontinent.

The peregrinations of Minhaj-us-Siraj Juzjani (1193–1260 CE) at the Ghurid court before he received his first appointment in the Delhi Sultanate by Sultan Iltutmish in the 1220s, and thereafter as a member of the ruling elite and the Sultanate court, included periodical religious assignments to garrison towns far from the capital, personal errands to and for family in Khurasan, travel emanating from his secondary interest in trading goods apart from his day job as a religious scholar, travel to find patronage as political authority changed hands from one Sultan or ruling house to another in Delhi, and travel of a more punitive sort, awarded due to his role in an apparent conspiracy or suspicion of disloyalty to the court (Kumar, 2007, p. 218). The array of contexts which culminated in the circumstances of travel for Juzjani belies any assumptions about the transitorial nature of travel and mobility. Additionally, it also suggests that essentializing the circumstance of travel and mobility in medieval India to individuals belonging (or rather not belonging) to particular socio-political groups or classes, or to certain phases of individuals' lives or states of the polity seems ahistorical.

The Persian Literati on the Move

Warlord polities dominated the medieval 'Islamicate'. In the same period in the Indian sub-continent, the advent of Turk nomad-tribesmen on horses had ushered in an era of conquest by horses. Patterns of conquest and consolidation of territorial sovereignty, especially by a state with sharper centralizing tendencies, necessitated mobility, and travel not just for conquerors but also for those constituting the court and administrative paraphernalia. The dependence of the state on the mobility of its political, religious, and literary elite tasked with the reproduction, propagation, and extension of state authority at various regional and sub-regional levels challenges the received wisdom on the sharp divisions between settled and nomadic societies. In this context, the medieval Persian literati, a subset of the *ahl-i-qalam* (people of the pen, which included administrators, secretaries, accountants, book-keepers, etc.) and a constituent of the ruling elite generally undertook necessitated departures.

When, in 1327, Sultan Muhammad-bin-Tughluq ordered a section of the ruling elite of Delhi to move to the 'new' capital at Daulatabad (Deogir), people screamed and cried while leaving their homes. Many in this forced political exodus to Daulatabad from Delhi were inflicted by what Isami called love for their place of birth. One such nonagenarian relative of our author, asleep in the confines of his home, was dispatched on the journey to Daulatabad in this unconscious state. Rudely awakened from his sleep, this nonagenarian, overcome by sadness, pangs of separation, and the travails of the journey, died midway (Rizvi, 2008, p. 99).

When members of the *ahl-i-qalam* such as Sadr al-Din Hasan Nizami (fl. 1205) were forced to immigrate to India owing to the political turmoil in Nishapur, he characterized his departure from his 'real residence' as a physical and spiritual separation, accompanied by hopeless and pale thoughts about life (Rasikh, 2020, p. 113). Yet another narrative on the struggles of leaving home and embarking on a journey came from the lexicographer Hajib-i-Khairat Dehlavi alias Ma'ruf who in his *Dastur-ul-Afazil* writes of his journey away from Delhi in the following poetic words:

> Since the revolving sky (cruel sky) selected me for companionship in wandering, it drove me away from my own real native city and the permanent residence i.e., Hazrat-i-Delhi and thus made me suffer from separation…As I did not possess the strength to resist, I took to wandering and set off on my journey. I moved from place to place, city to city, pouring tears from the eyes….'
>
> (Siddiqui, 1992, p. 83)

These mobilities accentuate certain themes which fall within the paradigmatic conceptualization of travel and mobility, just as they also undermine certain componentry elements of the same paradigm. The theme of departure is strongly emphasized. These instances of travel are necessarily geared toward destinations, and, to some extent, also arrivals. The life stories of these authors have also indicated in the hearts and minds of these men (and almost never women) a preference for sessility, 'attachment or fixation to one place' or a sessile condition, i.e., 'a continuous sameness of locale that intensifies a sense of temporal repetitions and may be escaped through departure', as opposed to their then-current state of mobility (Leed, 1991, pp. 4 & 17). It seems that the historiographical assumption of a separation between mobile individuals or communities which adopted travel and mobility as a lifestyle as opposed to sessile individuals or communities that preferred a 'stable' form of survival and sustenance *sans* the constancy of mobility was shaped by the words of authors like Nizami and Ma'ruf above.

My argument is that the aspirations of sessility by the *ahl-i-kalam* were as utopian as the idealism that pervaded the works of this Persian literati when writing about their patrons. Indeed, it was an aspiration precisely because sessility was a difficult prospect in the political context outlined above. Travel and mobility for our travel(ing) authors intensified further as independent Sultanates emerged in Malwa, Jaunpur, Kalpi, Bengal, and Gujarat. The presence of many regional sultanates offering patronage instead of a single imperial court, in what was a 'politically fragmented landscape', united in its vocabulary and assumptions, multiplied sources of patronage and opportunity, but also premised these opportunities on the willingness to travel (Alam & Subrahmanyam, 2009, p. 54). Just like members of the military and religious elite moved from one Sultanate to another in search of career prospects, so did writers, poets, and authors. Not only that, as socio-political developments rooted and provincialized elements of the Delhi Sultanate and the broader Islamicate (Patel & Leonard, 2012, p. 3) in the Indian subcontinent, the politics of the fifteenth century implicated a broader social configuration of the literate class; not just Persian but also those hitherto patronized in local kingdoms and writing in classicized Sanskrit and dialectical Hindavi.

Travel(ing) to Write: Shihab Ḥākim's Journey(s) from Jaunpur to Malwa

Shihab Ḥākim travelled to Shadiābād, the capital of the Malwa Sultans, in 870/1465–1466, at the age of 89. Here, he authored the *Ma'āṣir-i-MaḥmūdShāhī*, a *tawarikh* or historical text on the achievements of Sultan Maḥmūd Shāh Khalji (1436–1469), the most famous Sultan of the Malwa Sultanate and his patron. Ḥākim specifically mentions that he had always wanted to be at the Malwa court, but no such opportunity presented itself.[1] The role of the neighbouring Sultan of Jaunpur in delaying his visit to Malwa earlier, by refusing permission, is clearly stated (*rasīdan be pāyasarīr-i-'alā'bewaste mumā'anatay āmir jaunpur maisai name shud*). As Shihab Ḥākim does not name the Sultan, it becomes difficult to trace his whereabouts at different points in time. Also, now when he eventually managed to travel in 870/1465–1466, he credits it to none other than his avid wish, which had gained complete ascendancy over him and had brought him here to Malwa (*darīn waqt ishtiāq-i-khākbosi-i-dargāh-i-aʿlam-panāhbe-ḥaiṣīyati ātishzadan giraft*).

Shihāb Ḥakim was also the composer of a *masnawī* entitled, *urwatu'l-wuṣqá* (*ʿurwa'iwuṣqá*)², literally a strong or firm handle of faith, religion, etc. derived from a phraseology of the Quran. This work was composed in 859/1454–55 for an unnamed patron by an author with the pen name Shihābi (Shirani, 1966, p. 347). The contents of the work include ten chapters in the form of advice literature or the more commonly known mirror of princes genre, with chapters on Islamic law, justice, the art of governance, and generosity among others. While Shihāb Ḥakim's statements in the *Ma'āṣir* regarding his 'first' visit to Malwa in 870/1465–1466 would suggest that this earlier work was written for the Sharqi Sultans at Jaunpur as Iqtidar Husain Siddiqui speculates, it is likely a work composed by Shihāb Ḥakim for the Malwa Sultan, Mahmud Shah Khalji (1436–1469). Evidence to support this claim is hard to come by not only because the patron is unnamed but also because the essence of Shihab Ḥākim's *urwatu'l-wuṣqá* carries factual information of little import. However, among these meagre details is a reference to a royal order commissioning a palace by the name of Dhār-Mandu and the poet receiving ten villages and 81 horses from the Sultan in award (Shirani, 1966, p. 347). The probability of a palace named Dhār-Mandu, being commissioned by the Sharqi kings of Jaunpur, their competitors, is not high. This has led me to infer that the *urwatu'l-wuṣqá* was composed in Malwa.

Thus, two accounts of Shihab Ḥākim's life in Malwa are possible: the first, where we consider the *urwatu'l-wuṣqá*, which carries few actual facts and deals mostly with legendary curiosities and technological wonders of the *'aja'ib o gharā'ib* genre, as an *akhlāqi masnawī*, written in Jaunpur, either for the Sharqi Sultans or clandestinely for the Malwa Sultans. It may then have been the basis of Shihab Ḥākim's boasts that the Sultan had invited him to his court and had promised his patronage and that the Sultan of Jaunpur impeded his journey. On the other hand, this would require us to reject the information Ḥākim provides us about the award and the village grants he received as a fantasy of the author. It would also require us to reject the author's reference that he was ordered by the Sultan to work on the *masnawī*, and other granular details such as his visit to the unnamed Sultan's court on the eve of the Persian New Year (*nau-roz*) and the similarity of the Sultan's title in the two texts (*ḥaẓrat-aʿlam panāh khudāy-gānī*).

The other possibility, the one more likely to me, is that he had been in Malwa earlier (859), but he was not a permanent resident at Shadiābād and that he travelled. In regard to his journey in 870/1465–1466 described in the *Ma'āṣir*, we are told that he reached the capital, Shadiābād, and presented his works to the Sultan's son, who recommended him, but that men in the court were intent on soiling his reputation. He also writes that he wanted to come earlier but was not permitted by the Jaunpur Sultan. His presence in Malwa twice in two decades looks like two visits from Jaunpur, which was his probable permanent residence on both events.

That Jaunpur was his residence is also proven by references from a fifteenth-century lexicographer, Ibrāhīm Qiwām-al-Dīn Fārūqī, who was also originally from Jaunpur, had travelled to Bengal, and attended the court of Sultan Abu'l

Muẓaffar Bārbak Shah (r. 864–879/1460–1474). He quotes in his work frequent conversations and discussions he had with Amīr Shihābuddīn Ḥakim Kirmānī on various philological aspects of Persian vocabularies. These discussions, therefore, must have preceded the journey of these two experts in two different directions (Karamat, 2014, pp. 134 & 142). Also, for his visit to Malwa in 870/1465–1466, we are specifically told that in his zeal to come to Malwa, he turned his back on his *atfāl* (infants or boys, in plural form may also be used for wife), *aulād* (children), and *ahfād* (friends or relations), meaning that he travelled on his own. That his family was still in Jaunpur indicates that there was an intent to return, and this may have been, indeed, the condition of the unwilling Jaunpur Sultans to permit these periodical visits. What is certain in spite of this motley of contradictory facts is that Ḥakim had more than one place to be.

When Ḥakim was commissioned to write the *urwatu'l-wuṣqá*, he was 78 years old. He calls himself old and divulges little into his personal circumstance except dedicating a couplet to his son, Faizullah. When Ḥakim was commissioned to write the history of the Malwa Sultans, he was 89 years old. He calls himself old, weary, and troubled. We also know that he was travelling without the comfort and solace of family. Finally, he was travelling to and from Sultanates which were at constant loggerheads. The relationship between the Malwa and the Jaunpur Sultans was icy, owing to the claims of the two sultanates over Kalpi. With relations to maintain in Jaunpur and Malwa, writing about his two patrons who had been in conflict over Kālpī was difficult. What is also certain is that Ḥakim had more than one place to be. In this context, how could he best write about the complicated relationship of his present and past (and presumably future) patron?

Works produced within the inter-sultanate political matrix, mostly court chronicles like Hakim's, tended to be dry reports with little narrative or personal experience and observation making it into the final work. While a number of reasons, including the author's personal style and preference, availability of information, and the patron's command may have affected the nature and form of the final work, it seems plausible to suggest that inter-sultanate mobility, which may have grown out of choice or search for livelihoods, led many authors to turn into travel(ing) authors. Concomitantly, they were forced to choose reticence in narrating episodes which they had seen or experienced from close quarters or were sometimes themselves involved in, leaving their texts dry and devoid of any analytical flair.

Hakim's recording of contemporary events, which he probably witnessed as a member of the Malwa or Jaunpur court, such as the arrival of an ambassador from the Delhi court seeking the Malwa Sultan's help against Mahmud Sharqi's attack on Delhi in 1467, gives little hint of his obvious familiarity with the Jaunpur Sultans. He writes of the matter as an outsider to the events; however, considering the fact that he was patronized at Jaunpur, he was bound to not only know a lot more but also to have had an opinion about these events. His criticism of the Jaunpur Sultan on behalf of the Malwa Sultans is also extremely measured. This comes across clearly when we read his narrative of events involving other (that is, not the Jaunpur Sultans) competing rulers, who were assumedly not part of his travelling itinerary, where he provides an elaborate account of the ruler's cruelty, painting him as a tyrant and torturer. It was not possible for our author to apply the same standards of awareness or exaggeration to the Sultan of Jaunpur.

So, even though, possessing both sides of the story of a contemporary event, our author reduces the account to a plain narration of events which would do justice to his present patron's sentiment without reflecting any personal involvement or information with the events being described. It is merely a report.

The limited nature of certain fifteenth-century texts, where authors wrote merely what was reported to them and curated little space for their own observation or perspective on contemporary politics, suggests that they chose to be compilers of history and not authors per se. It seems that as subordinate partakers of the politics of the fifteenth century, and as men who travelled to and found patronage in different sultanates, some of which were in a permanent state of conflict with each other, an indiscreet voicing of one's views was impolitic.

As a result, when historians attempt to find the analysis of characters which Ziauddin Barani (fourteenth century) excelled at, or the short but insightful and authoritative details that Minhaj-us-Siraj Juzjani (thirteenth century) was able to append to his accounts of the slave Sultans, they are often left disappointed. But the fact is that neither like Barani did Ḥākim have a lifetime of experience of serving at one court, nor like Juzjani was he a permanent appointee who had sustained his position of favour through multiple reigns by aligning with prominent members of the ruling elite, the result of which was the biographies of these men in his (Juzjani's) work. While Barani and Juzjani spent a lifetime in Delhi, Ḥākim travelled for work.

It is important to underline that while the personal circumstances of our authors may differ, our focus here remains on the commonality of the 'travel for work' experience for authors and poets in the fifteenth century, which emerged from the existence of multiple, conflicting sources of livelihood which could be availed only with a willingness to travel. This travel could involve movements from one court to another or from one sultanate to another, but such was the prevalence of the idiom of 'travel for work' in this period that our roving authors provide evidence of transcending not just political borders but also socio-politically distinct, regional frontiers. The works and activities of poet Māhēśa under the tutelage of Rāṇā Kuṃbha in Mewar (c. 1433–1460) indicate that the 'travel for work' idiom was not restricted to the Sultanates of the fifteenth century. Indeed, the proliferation of these regional, Indo-Islamic polities in the fifteenth century meant that multiple sources of patronage were available not only for authors in Persian, the classical, administrative language of the Delhi Sultans of the thirteenth-fourteenth centuries, but also for authors in Sanskrit, the classical language of polities outside of the Sultanates of the fifteenth century.

Travel(ing) to Write: The Peregrinations of *Kavi* Māhēśa in Mewar and Malwa

Kavi Māhēśa came from a long line of *dashora* Brahmans patronized by Rāṇā Kuṃbha as royal poets, commissioned to compose elaborate Sanskrit eulogies in the honour of the Guhila lineage. These eulogies have been inscribed on massive stone inscriptions found at the Kumbhalgarh Fort, the Chittorgarh Kirtistambh, and the Eklingji temple complex near Udaipur.

Māhēśa records the encounters between Rānā Kumbha and the Malwa Sultans, primarily Sultan Mahmud Khalji of Malwa, in two inscriptions dated 1517/1461, i.e., the Kumbhalgarh and Chittorgarh Kirtistambh inscriptions (Sarda, 1917, pp. 16–17) and the Eklingji Mahatmaya inscription dated 1545/1489 (Peterson, 1899, pp. 117–133). The picture drawn in these inscriptions presents Kumbha as an invincible warrior, supreme over all his contemporaries. We are told that Kumbha 'placed his left foot on the head of the brave lord of Mālava and took Janakachal'. In another such encounter, 'he humbled the pride of the Muhammadans at Sārangpur and imprisoned several young women of the lord of the parasikas'. Also, 'he defeated the ruler of Vrindavati (Bundi)…captured the fortress of Gangaratta (Gagraun)' (Sarda, 1917, pp. 180–181). The second inscription of the same date informs that:

> He… conquered the whole kingdom of Mudafar (Muzaffar Shah)…He subjugated Muhammad, destroyed the town of Nagapur with the lofty mashiti, built by Peroj, imprisoned several young Muhammadan women and took possession of the treasure of Shams Khan…He destroyed the common armies and the lords of Malwa and Gujarat in war and captured several elephants.
>
> (Sarda, 1917, pp. 183–185)

The authenticity of these events is a matter of debate, but our focus here is on the poet's awareness of the regional and political context in which he had to build his patron's supremacy.

Sometime after 1517/1461, Māhēśa had travelled to Khadaoda on the Malwa-Mewar border, where he wrote a sixty-nine-verse-long eulogy in 1485 for a new patron, a convert to the religion of Islam, Malik Bahri, who recognized the tutelage of the Malwa Sultans at Khadaoda. Mahesa was an encomiast for the same Malwa Sultans whom he had derided till now. In this inscription, located in a new spatial context, the poet states that 'Śaṅkara and Kārtikēyḥ left their heavenly abode at Kailāśḥ to reside in Malwa ', that 'worthy men are not fearful of coming to Malwa', and that the rich of this city (Mandu) are not wicked or deceitful. This, incidentally, also suggests that Māhēśa had not limited his visit to Khadaoda on the Mewar-Malwa border but had visited the capital, if only for a brief period.[3] He also celebrates Bahri's generosity in the 29th verse (Ojha, 1931, p. 1).

Four years after the 1485 Khadaoda inscription, we find Māhēśa back in the service of the Mewar Rana, Raemall, who had managed to gain an upper hand in the post-Kumbha succession struggle which lasted for six years (1530/1474). Back now in Mewar, Māhēśa received the grant of the village Ratanakhera from the Rana, and wrote a hundred-verse-long eulogy for his new patron in 1489.

In an instructive analysis of the mental workings of these writers, Teuscher states that the job of producing inscriptions (or texts like the *tawarikh* and *tabaqat* for that matter) in Mewar required 'research, historiography, problem-solving, drafting, re-drafting and adjusting the gathered facts according to a highly complex and very visible code of meaning which constituted a large part of the communication between elites within a given area' (Teuscher, 2013, p. 181). Imagine the complexity of this task when this very visible code of meaning which tied the kingdom to its elite through a carefully curated historical and contemporaneous narrative of the kingdom and its kings vis-à-vis others had to be aligned by the author, also to his own mobility and, concomitantly, changing clientele!

This is probably why the encomiums that poet Māhēśa strung together for Rana Kumbha's nemesis, Mahmud Khalji, at Malwa, reflect an attempt by the poet to certainly build a glorious legacy for the Malwa Sultan, but one that was separate and did not implicate the Rana of Mewar or the sanctity of his authorship of the encomiums dedicated to the Rana of Mewar. Relating the Sultan's victorious campaigns from the 15th verse onwards in the Khadaoda inscription, he writes that right at the beginning of his *digvijaya*, the Sultan (i.e., Mahmud Khalji of Malwa) raised the din of his power in Delhi. His terror went far south to the Chōla kingdom, caused the drubbing and destruction of Utkala, and brought great trouble to the formidable Drāviḍa kings (Ojha, 1931, p. 29, v. 15). He also writes that Salah, Bahri's political master who resided at the Malwa court, had killed 80 elephants of the Gujarat ruler on behalf of Mahmud Khalji. It is difficult to determine where the poet's list of Mahmud Khalji's victorious campaigns came from, but it seriously under-represents the Sultan's military activities. If the beginning of the *digvijaya* can be taken to mean the beginning of the Sultan's reign, there is some truth in his first statement, which refers to the Sultan's quasi-military campaign to Delhi (Day, 1965, pp. 116–119). The last piece of information regarding the killing of 80 elephants of the Gujarat Sultan do not appear in the *Ma'āṣir,* though a generic reference to war booty does. Another near-contemporary text from the neighbouring Sultanate of Gujarat, the *Ta'rīkh-i-Mahmūd Shāhī,* on the other hand, gives contrasting details about the same events, mentioning at one point that Sultan Qutbuddin of the Gujarat Sultanate had acquired 80 elephants of the Malwa army during this battle (Tuni, 1988, p. 70). It seems that poet Māhēśa was highlighting Salah's role in bringing back these elephants to the Malwa Sultan. However, the realistic vein of the poet's eulogy ends here. The Sultan's own court-chronicler, i.e., Shihab Ḥākim who was hired with the express intent to record the Sultan's achievements for eternity, makes no reference to Utkala, the Chōla rulers, or the Drāvida kings or the medieval equivalents to those regions. Neither does any other contemporary court-chronicler make any mention of the three campaigns or victories which Māhēśa refers to. It is also obvious that kingdoms like Chōla and Drāvida were temporally irrelevant to the fifteenth-century Malwa Sultanate. We also know of no expedition of Mahmud Khalji to Utkala.

What Māhēśa seems to have done, therefore, is conjure a victorious military career for Mahmud Khalji out of stock phrases and literary tropes. The insertion of these stock phrases, which assisted in the manufacture of grand encomiums or varnish over uncomfortable truths, would have made sense if the Malwa Sultan had spent a lifetime at the capital and had little in terms of military achievement to boast about. But this was simply not true. The reputation of Mahmud Khalji as a warrior king was established in his lifetime by a long and continuous list of arduous military campaigns, which covered Gujarat, the Bahamanīs of the Deccan, the Sharqī, and the Lodī Sultans in the North.[4] One may especially note the Sultan's recorded campaigns and occupation of regions in and near Mewar, especially Mandasor (858–859), Mandalgarh (846–847), Kotah (851), Gagraun (846), Ajmer (860), and his attacks on Chittorgarh (861), Kumbhalgarh (863), and Ranthambore (850), all of which fanned the flames of enmity between the two contiguous Sultanates.[5] One particularly feels the ahistorical vein of our poet's eulogy to Mahmud Khalji when compared with his own description of Mahmud Khalji's predecessor, Sultan Hoshang Shah's (1410–1436) achievements wherein he mentions his most prominent and well-known victories at Jajnagar and Kālpī. That he could not do so for Sultan Mahmud Khalji, information about whose campaigns did not require digging through old material, and may have been experienced by the poet as a court chronicler at Mewar suggests that it was a conscious act of omission and discarding available information on the part of the poet.

It is important to understand why Māhēśa chose to obliterate the Malwa Sultan's achievements, especially those related to Mewar and its vicinity. However, all conclusions can only be tentative because, unlike textual sources, our knowledge of the reach of these inscriptions, i.e., the extent of the regions to which the inscribed information reached, and the size and nature of the audience which formed its readership, is poor. Nonetheless, the distance between the two places (about 180 kms. today) where Māhēśa composed these inscriptions was not so much as to ensure an entirely new audience for him. An admission that vast swathes of Mewar had fallen into the hands of the Malwa Sultans would be difficult to resolve with the *digvijaya* he had ascribed to the Rānā in Mewar and hurt the poet's credibility with the local elite. While the earlier text had already been inscribed and had found an audience, it was the content of the Khadaoda inscription which had to be altered to sit in harmony with his past works while ensuring that the demands of his present patron were not sidelined either. Thus, the idea of a *digvijaya* stays but the regions are substituted.

One must, however, not misunderstand Māhēśa's choices as a question of loyalty, political or religious. He wrote about the defeat of Kshemkarna, a brother of Rana Kuṃbha, at the hands of Bahri, his patron, while consistently referring to him as a *yavana* (Ojha, 1931, p. 44, v. 43–44). What we find here, on the contrary, is a perceptible sense of spatial awareness of the narrative he manufactures and communicates and the audience which consumes it.

The same discretion guided Māhēśa at the next stage of his career at Ratanakhera where from his 100-verse-long eulogy one learns of the carrying off forcibly of 'a large fine from Malava after tearing up the trees around Kherawada, and cutting numerous *yavana* into pieces with the sword, and uprooting the families of his enemies' (Ojha, 1931, pp. 43–46, v. 26 & 35). These *yavanas* must have included Malik Bahri, his patron of four years ago. What Māhēśa does not tell us is that he most probably had also returned or was forced to return to Mewar as a consequence of this invasion by the Rānā on the north-western border of Malwa.

Teuscher has argued that while the king ordained the inscription to communicate the message of his sovereignty, it was often left to the composers to bring out the ruler's supremacy in the knowledge and content framework which had been used for generations and in which they were specialists. Māhēśa's ability to tweak historical details to resolve the requests of patrons by inserting a new set of historical or ahistorical details was an example of the author employing his repertoire of skills to reconcile the conflicting claims of his present, past, and future patrons while ensuring an unsullied reputation and a career across journeys.

One does not have enough insight into the development of our poet's professional skills and their relation, if any, with the travels and travails of his career. It is difficult to say, for example, if these literary tools contributed to Māhēśa's amenability to travel and deliver the expectations of different patrons in distinct politico-spatial contexts, or whether the embedded geographical mobility in medieval India, in the form of necessitated departures, shrinking, and expanding sources of livelihood, or the increasing permeability of border regions in the fifteenth century, pushed poets like Māhēśa and Shihab Hakim to tackle the new challenges which emerged out of travel(ing) for work by expanding their repertoire of writing and communication skills.

Travel(ing) to Write: A Fifteenth-century Phenomenon

What is clear, nonetheless, is that mobility was the foremost trait which defined the lives of authors and writers in this period. Just like Hakim and Māhēśa, the careers and writings of many other authors in this period were determined by mobility and amenability to travel. Abdul Karim Nimdihi came from Laristan and Shiraz, travelled to Mandu in the hope of employment but failed to find a job there and eventually wrote for the Bahamani and the Gujarat Sultans later. The life of Qāzī Badruddīn Dhārvāl Dehlawī, a fifteenth-century lexicographer, was similar as he travelled from Delhi to Jaunpur to Malwa (Dhar) and then in the times of Qadr Khan (c.1405–20) to Chanderi, giving us a glimpse into the frequently shifting patrons of these writers (Akhtar, 1995, pp. 358–359).

The author of the *Ta'rīkh-i-Mahmūd Shāhī*, Abdul Ḥusain bin Hājī Tūnī, was a resident of Tun, a village in Khurasan, where he lived with his family while his father was serving in the Deccan (Tuni, 1988, p. X). He studied at Isfahan and also spent a considerable part of his mature age in that region, before travelling to Gujarat for his assignment. This shows in his work on the Gujarat Sultanate where he fits multiple Sultanates within his wider canvas. He chose an annalistic form of history writing whereby he used year-by-year accounts for multiple Sultanates such as the Gujarat, Bahamanī, Malwa, Delhi, Jaunpur, and, keeping alive his connection with his native place, also the Timurid Sultanate in Iraq and Khurasan. However, the consequence of this wider context was that its 'wearying, fulsome panegyrics' had 'no potential for detailed causal analysis or even a systematic in-depth study of a single region', something which was at least attempted by Shihab Ḥākim for Malwa (Tuni, 1988, pp. V & IX).

Our discussion above of the lives of these fifteenth-century roving authors shows that with the demise of the Sultanate, opportunities grew as political and social redlines were defied and politics became more intrusive. Sources of patronage were no longer centralized, and courts were spread across north India. Travelling for work opportunities also grew concomitantly. Medieval mobility was pervasive anyway, but the circumstances of the fifteenth century seem to have further expanded the fold of travellers to include travelling writers, further enriching the firmament of medieval mobilities with the lived experiences of individuals typically conditioned as sessile and stable. How must, then, one explain the aspiration for sessility in the context of these pervasive, embedded mobilities?

Partially rejecting the characterization of these aspirations as mere literary tropes, I suggest that any contrast between mobility and sessility in the context of the medieval Indian subcontinent, especially in the Persian literati's desire for one as opposed to the other, should be constructed as an anxiety for constancy of livelihood sources and patronage opportunities. We argue through the cases of the careers of our roving authors that, especially in the fifteenth-century northern Indian political landscape, one finds evidence of travel, migration, and mobility as both disruptive forces pushing travel(ing) authors out of their comfort zones and also as opportunities to build and exercise agency. Travel and amenability to travel accorded opportunities and while the theme of separation from home and nostalgia for that which was left behind pervades the works of most of our authors, these new patterns of mobility and patronage also reveal uniquely indigenous responses to their new socio-political environs. Roving authors may have utilized the concept of home and away in their works as opportunities to inscribe and exaggerate the self by shining light on their troubles through these biographical notes, and therefore justifying requests for patronage, but they simultaneously seem to have also embraced these intensifying mobilities by developing a repertoire of innovative strategies to acquire and narrativize information and to find work and establish their identity in varied contexts. Finally, this also informs our perspective on medieval peregrinations which come across as a socio-politically transcendental phenomena, which impacted by the fifteenth-century political context seems to have become less transitorial, more routinized and quotidian, thus preparing our court chroniclers for a life of travelling and writing as travel(ing) writers.

Notes

1 The information here comes from N.H. Ansari's preface of the edited Bodleian Library manuscript of the Maasi'r-i-Mahmudshahi, published in New Delhi in 1968. An unedited copy of another manuscript of the Maasi'r-i-Mahmudshahi at the Tübingen Library, Germany, has been digitized by the Berlin State Library, identified as MS. or fol. 535 is available online. The author's preface in this manuscript, of approx. 137–138 folios or 265 sheets, runs into the first nine folios. In this extremely densely written preface, there are few references to the author's life to corroborate the information found in N.H. Ansari's preface. As I have been unable to read and comprehend the Persian text in its entirety, I have relied entirely on N.H. Ansari's edited text, especially his preface, to glean details of the author's life. The corroboration of the information regarding the author's life given in the editor's preface to the edited text with the unedited Berlin Library manuscript remains a desideratum.
2 The masnawi is not available in its manuscript form to the best of my knowledge. An extremely valuable essay by Mahmud Shirani in the sixth volume of an essay collection

entitled *Maqalat-i-Mahmud Shirani* provides an excerpt of the couplets in the masnawi, a single page of the manuscript form, and an original commentary on its contents. These are the basis of the information above. However, excerpts produced by Shirani only pertain to the creations of the author (Hakim), for instance, the commission of a pavilion which harmoniously combined natural elements like fire, water, and air, or a clock that chimed on Hindustani time-keeping patterns. These were the marvels of Hakim's age. The inadequacy of the reproduced section comes across well when we consider that of the extant 3500 couplets of the masnawi, Shirani has reproduced less than half.

3 Mahesh's statement that he was famous for some time in Malavadesa due to his *kavya* would also suggest that the journey and time he spent in Malwa was planned by him for a brief time period and reflected a temporary arrangement.

4 While there were multiple campaigns against each of these polities, the prominent ones were: the Battle for Champaner, Kaparbanj, and Surat along with a final treaty between 852–855/1448–1451 (Day, Medieval Malwa, pp. 124–125 and *Maasi'r*, pp. 68–73); Kalpi-Jaunpur struggle reached its acme in 848–849 and was followed by another political agreement (Day, pp. 140–145 and Maasi'r, pp. 61–62); the Bahamanis were met but not defeated in various conflicts across their southern border (Day, pp. 147–155); etc.

5 Hakim, *Maasi'r*, pp. 80, 51–52, 65, 54, 83–84, 85, 89, and 64, respectively, dates are given in Hijri calendar and differ slightly from Day and later secondary works.

References

Akhtar, M. S. (1995). Dharval, Qazi Khan Badr Mohammad Dehlavi. *EncyclopædiaIranica*, *7*, 358–359.

Alam, M., & Subrahmanyam, S. (2009). *Indo – Persian travel in the age of discoveries, 1400–1800.* Cambridge University Press.

Day, U. (1965). *Medieval Malwa: A political and cultural history, 1401–1562.* MunshiramManoharlal.

Hakim, S. (1968). *Maasi'r-i-Mahmudshahi* (N. H. Ansari, Ed.). Idarah-i-Adabiyat-i-Delli.

Karamat, D. (2014). Turki and Hindavi in the world of Persian: Fourteenth and fifteenth century dictionaries. In S. Sheikh & F. Orsini (Eds.), *After Timur left: Culture and circulation in fifteenth century North India* (pp. 130–165). Oxford University Press.

Kumar, S. (2007). *The emergence of the Delhi Sultanate, 1192–1286.* Permanent Black.

Leed, E. J. (1991). *The mind of the traveler: From gilgamesh to global tourism.* Basic Books.

Ojha, R. S. (1931/1988). Indore Museum ka ek Shilalekh. *Nagari Pracarni Patrika – 1988*, *12*, 1–96.

Patel, A. & Leonard, K. (Eds.). (2012). *Indo – Muslim cultures in transition.* Brill.

Peterson, P. (1899). *A collection of Prakrit and Sanskrit inscriptions.* Bhavnagar Archaeological Department.

Rasikh, J. S. (2020). The many lives of a Medieval Muslim Scholar: An introduction to the life and times of Minhaj Siraj al-Din Juzjani, 1193–1260 CE. *Afghanistan, 3*(2), 111–134.

Rizvi, S. A. A. (2008). *Tughluq Kaleen Bharat – Part I.* Raj Kamal Prakashan.

Sarda, H. B. (1917). *Maharana Kumbha: Sovereign, soldier, scholar.* Scottish Mission Industries Ltd.

Siddiqui, I. H. (1992). *Perso-Arabic sources of information on the life and conditions in the Sultanate of Delhi.* Centre of Advanced Study in History Muslim University.

Shirani, M. M. (Ed.). (1966). *Maqalat-i-Hafiz Mahmud Shirani* (pp. 341–406).

Teuscher, U. (2013). Craftsmen of legitimation: Creating Sanskrit genealogies in the tenth to fifteenth centuries. *Afghanistan, 29*(2), 159–182.

Tuni, A. H. (1988). *Tarikh-i-Mahmud Shahi* (S.C. Misra, Ed.). Baroda Oriental House.

5   The Power of Itinerancy

Religious Leaders in the Nepal-India Borderland

Martin Gaenszle

Most renouncers in South Asia, such as Shaiva or Vaishnava ascetics, Sant gurus, bhaktas, etc., traditionally follow an itinerant lifestyle and thus have always tended to be ignorant of state borders as they travel in a different kind of space (e.g., Burghart, 1983[1]). They can be seen as a particular type of 'trans-border individuals'.[2] It is well known that in South Asian history these religious figures, due to their special status and privileges, have often been used to play the role of spies and informants. Bayly stresses, for example, the role of Gosains as an 'intelligence community' providing information for military use by the rulers (Bayly, 1996, p. 69). The activities of ascetics as soldiers, power brokers, and 'yogi spies' are well documented (Pinch, 2006; White, 2009, pp. 220–231). But apart from 'fake sadhus' (who deliberately disguise themselves as ascetics to hide their identity), there are also those itinerant religious leaders who have a genuine cause of faith and some other agendas which they propagate while moving from village to village or town to town. In fact, the role of such wandering ascetics included the dissemination of news from the larger world, i.e., they fulfilled the function of providing a communication network.

This chapter looks into the lives of two important individuals in the late nineteenth and early twentieth centuries who began their careers in East Nepal and – even though the Ranas, who ruled the country with an iron hand, kept a strict border regime – frequently moved across the border into Darjeeling and Sikkim and Northeast India in order to propagate their creed and gather a following. This borderland around the Singalila Ridge was an important contact zone between Nepal and British India. Though for foreigners, this border was difficult to cross, the locals moved rather freely – in both directions. Thus, the itinerant religious leaders, who both maintained certain benevolent links to the forces in power, also managed to evade the political control of the state and even went against it in acts of resistance.

One of the two individuals is an important guru of the Josmani sect, Gyan Dil Das (1821–1884): he was a Brahman from East Nepal who spread new bhakti ideas, especially among Rai and Limbu villagers, and eventually had to leave the country for political reasons and settled in Sikkim, along with his disciples. Even today, there are small communities of his followers who reside in the Geling area of Southwest Sikkim.

The other is Phalgunanda Lingden (1885–1949), the Limbu founder of the so-called Satyahangma Dharma, a new religion which resulted from a fundamental

DOI: 10.4324/9781003544524-8

reform of the Limbu's ethnic religion in the 1920s. It is mainly popular among urban Limbu in Nepal but also has a good number of followers in India. Phalgunanda is today regarded as one of the national 'luminaries' who have been acknowledged by the government.[3] Yet during his lifetime he was often seen as a seditionist and even prosecuted by the Rana state. In his youth he travelled a lot, even reached Europe, and wandered throughout India, but later he clearly focused his proselytizing activities on Nepal.

Both religious traditions (or 'sects') can be seen as 'trans-border religions',[4] i.e., they operated by transcending national boundaries, and both played an important role in the Singalila borderland. In fact, as I will try to show, both traditions benefited in crucial ways from this cross-border interaction. Moreover, both traditions, though they remain distinct, have become closely connected historically during the last century, as both strongly affected the development of Kirat Dharma, i.e., the religion of the Kirat people as it is practised today (consisting mainly of Rai and Limbu ethnic communities).

This relatively new religion – Kirat Dharma – is based on ancient indigenous traditions, but it has been partly influenced by the reform movements of both the Josmani as well as the Satyahangma. It was acknowledged as a distinct creed first in the Census of 1991,[5] but one must stress that this religion is still in the process of self-definition; it is still in the making. Therefore, the history of the two traditions under discussion has to be seen in the larger context of current debates on what constitutes an authentic and, at the same time, modern Kirat religion. These debates have been going on in Nepal, Sikkim, and Darjeeling as part of ethnic politics, as ethnic identities had to be continuously defined vis-à-vis the state authorities.

The significance of boundaries and 'trans-border' mobilities for the process of redefining Kirat religion comes out clearly in a recent controversy over the role of 'Yumaism' in the constitution of Kirat Dharma. The concept of Yumaism has been particularly popular in Sikkim and is propagated by Jash Raj Subba, an influential intellectual and writer (cf. Vandenhelsken, 2021). It has recently become an issue among the Limbu in Nepal, who partly welcome it, but there are also others who feel that the idea of Yumaism divides the Kirat people, separating the Limbu from the larger religion. This position has been forcefully made by Manjul Yakthumba in his book on '*Ko hun Kirāt ra Kirāt-Limbū? Ke ho Yumādharma?*' (2018).

Yumaism is a kind of reformed Limbu religion which emerged in Sikkim and has been gaining popularity in recent years. Yumaism has been argued (Gustavsson, 2013, p. 93) to have been considerably influenced by the Satyahangma movement (i.e., the ideas of Phalgunanda), but due to the different political situation in Sikkim, where the Limbu have a prominent position as a major indigenous community, it developed as a distinct marker of Limbu identity, different from other ethnic groups. Both religious traditions focus on the worship of divinities at temples (*manghim*) and propagate a pure lifestyle. At the same time, they both stress the ancestral ethnic heritage. However, the Satyahangma followers as well as traditionalists in Nepal vehemently insist on a common Kirat identity of all Kirat groups. Here the emphasis is more on the overall cultural relatedness of all Kirat groups, not only the Limbu (or, as they prefer to call themselves, Yakthumba) but also including the various Rai groups, the Yakkha, and Sunwar.

In other words, the different social and political backgrounds of Kirat communities in Nepal and India has led to diverging traditions, hence a crisis of self-identity and search for cultural roots. This necessitates a critical look at the historical processes that affected their development. Clearly, the role of highly mobile itinerant religious leaders was crucial for the emergence of new identities: making use of an established tradition of vernacular mobility, these leaders easily crossed national boundaries and evaded, when necessary, the dominant powers of the state, be it the autocratic Rana regime or the colonial British Raj.

Gyan Dil Das

The Josmani are a bhakti sect (*paramparā*) of the Sant tradition that worships the divine as *nirguṇa* (formless). It was once very influential in Nepal during the late eighteenth century and much of the subsequent century, when it was highly esteemed even in aristocratic circles, among both the Shah royals and the Ranas. But, for reasons outlined below, it fell into disgrace in the late nineteenth century and today survives in some pockets in east Nepal but mainly outside the country, in Sikkim (India).

The origins of the Josmani are not entirely clear, nor is the derivation of the name. The best source is still the book by Janaklal Sharma (1963), which contains a short history and a rich body of songs and writings (also see Timalsina, 2010). More recently, a field-based study on contemporary vestiges of the tradition was published by Novel Kishor Rai and Netramani Rai (2013). There is indication that the tradition was originally linked to Joshimath (near Badrinath, Garhwal) – which could explain the name. In any case, the Josmani's influence clearly spread from western Nepal to eastern Nepal and then beyond.

One early important guru was Shashidhar Das (ca. 1747–1849), who was seminal in Western Nepal. He was a contemporary of Prithvi Narayan Shah and later became the teacher of Rana Bahadur Shah, whom he initiated into the order (Sharma, 1963, pp. 65–67). Shashidhar is also said to have initiated the proselytization of the creed: as the story goes, he had four disciples from different ethnic communities (*jāti*): one Magar, one Gurung, one Sunuwar, and a Brahman. He sent them in the four directions to disseminate the tradition. Thereafter, and for most of the nineteenth century, the Josmani were well respected by the state.

But this changed after the death of Jang Bahadur Rana (in the 1870s). The reasons are not fully known, but it seems it had to do with the fact that the rebel king in Magar country, Lakhan Thapa, was a member of the Mokshamandal, the order propounded by one of the early disciples of Sashidhar, who was a Magar (see above). Lakhan Thapa had attempted a coup against the Rana leadership: he wanted to establish a Magar Kingdom, and, in fact, he was accused of attempting to assassinate Jang Bahadur. But the coup was forestalled – rumours had reached the army – and Lakhan Thapa was finally hanged in front of his house near Manakamana in Gorkha district (for details, see Lecomte-Tilouine, 2002).

It seems at least partly due to this that the fortunes of the Josmani changed in the late nineteenth century. This is indicated in the biography of the second most important guru, Gyan Dil Das (1821–1884), a highly prolific and influential teacher. Though he was a Brahman, he had a following that included men and women of diverse socio-economic and political persuasions and is known to have initiated many disciples in the villages of eastern Nepal, in particular among Rai and Limbu.

Like in other bhakti traditions in South Asia, the Josmani order, though originally constituted by Brahmans, opened up to lower castes and tribal groups, as its religious outlook was rather egalitarian and the gurus were against the caste system. This religious inclusiveness was attractive to the members of ethnic minorities, especially the local elites, as it provided a way to identify as Hindu in a Hindu state. And, as membership among the Josmani was transmitted from the father to the children, the ideas and practices of this tradition spread quickly all over the eastern districts of Nepal.

Gyan Dil Das was born in east Nepal as Shrikrishna Lamichane (Upadhyaya Brahman). There is some controversy regarding the exact location of his birthplace. According to Sharma, Das was born in Dhankuta district, but later research indicates Ilam district (Rai & Rai, 2013). He had received a Sanskrit education but became critical of Brahmanvad. At a young age, he was initiated into the Josmani *dharma* in Panchthar and was eventually sent to Rumjatar (Okhaldhunga), where he converted many Gurungs to the creed. There he married, and his son and disciple Ravidil Das later stayed in Rumjatar after his father's death. There is a statue in Rumjatar which has been erected to commemorate Ravidil Das.

Sometime in the 1870s, Gyan Dil Das also came to Kathmandu, but this was at the time of Lakhan Thapa's rebellion, and soon he was jailed for six months, watched over by Jang Bahadur himself. But Jang Bahadur, the story goes, was so impressed by the Guru that he let him go. Other prominent Ranas were initiated by the ascetic, namely, Ranauddip, the Prime Minister's brother. Gyan Dil Das eventually received official permission from Jang Bahadur to spread the Josmani tradition and returned to Rumjatar.

But there, soon after the death of Jang Bahadur (in 1877), he ran into trouble as the local district governor (*baḍāhākim*) banned his activities. Following this, Gyan Dil went east and eventually crossed into British territory where he was safe from prosecution. He first went to Gundri Bazaar, the main market in Darjeeling. He stayed there with a few followers and continued to spread his dharma. This brought him into conflict with the Christian missionaries who had begun their own proselytization.

There is an interesting story about this conflict between the Josmani and the Christian proselytizers in Darjeeling (see Sharma, 1963, pp. 70–72). It recounts how Gyan Dil Das was countered by Reverend Archibald Turnbull from the Foreign Mission of the Church of Scotland, which arrived in the 1870s in Kalimpong (Jayeeta Sharma, 2017). Turnbull tried to stop Gyan Dil Das's activities and threatened to drive him out of Darjeeling if he refused to cease propagation in the region. Gyan Dil, in return, challenged Rev. Turnbull to hold a competitive debate: whoever lost the debate would have to burn his book. Turnbull agreed, and after fixing time and place the debate was held in front of a large public audience. It went on and lasted for days. Eventually, Turnbull, so it is said, lost and had to burn the Bible! The ashes were mixed with honey into little balls, and these are still worn by the local Josmani today. It seems that Rev. Turnbull was a generous loser. He is even believed to have asked the British Government to give some land as *birtā* (i.e., freehold) tenure to Gyan Dil Das so he could build a *kuṭī* (hut). This was established in Rangbul, a place between Jorbangla and Sonada, on the road from Darjeeling to Siliguri. The Guru stayed there and gathered a large following of initiates around him. Moreover, during his stay in Rangbul, Gyan Dil Das also wrote important literary works, such as the *Udaya Laharī*.

Not much is known of the Guru's personal life, but it is reported that apart from his son (see above), he also had a daughter who had joined him and his wife in Darjeeling, whom he loved very much. When she suddenly died in Rangbul, he built a tomb (*samādhi*) for her there, which is now a pilgrimage shrine visited by Josmani followers (Sharma, 1963, p. 105).

In the course of time, some of his disciples (who came from various ethnic groups, such as Gurung, Rai, as well as upper Hindu castes) went on to build *kuṭi*s and *maṭh*s (monasteries). Thus, Nirvanadas Rai built a *maṭh* in Geling (West Sikkim) and invited Gyan Dil to stay there, which he did. Later, he sent Nirvanadas to another place in Sikkim, namely, Samdong (See Fig. 5.1 below), where another *maṭh* was built (which is called Rami Dham).

However, the main centre of Josmani proselytization was in the area around Geling in West Sikkim. This area, Chyakhung (or Chyakum), was under the control of a certain Jerung Devan. He was suspicious of the activities of Gyan Dil, as he considered the state a Buddhist kingdom and did not find it proper to have a Hindu missionary on his land. So, one day he went to Geling and destroyed the *maṭh*, trying to evict Gyan Dil Das from the land, which created quite an uproar. Eventually, some well-situated supporters arranged for a deal. Gyan Dil could stay under the condition that he discontinued his proselytizing activities among the Bhote and Lepcha communities. He agreed to this and went back to Geling where he died in the year 1884. His *samādhi* is still worshipped there. The Josmani creed further spread east, with followers in Bhutan and Assam and even Tripura, Manipur and Burma (Sharma, 1963; see Fig. 5.2).

What does this story of wandering and dissemination tell us? I think this is an interesting case illustrating how different political conditions in the borderland area were limiting conditions for religious communities yet enabling conditions for successful missionary activities. The mobility of the itinerant Guru allowed him to evade suppressive statehood and settle in a more tolerant environment. While Gyan Dil Das made effective use of these options, another guru from the same region, active about 50 years later, chose a different path.

Phalgunanda (1885–1949)

The Limbu religious leader and founder of the Satyahangma reform movement, Phalgunanda, comes from the same area as Gyan Dil Das, the Ilam-Panchthar region (Nepal) in the Singalila borderland. Though the two Gurus were not

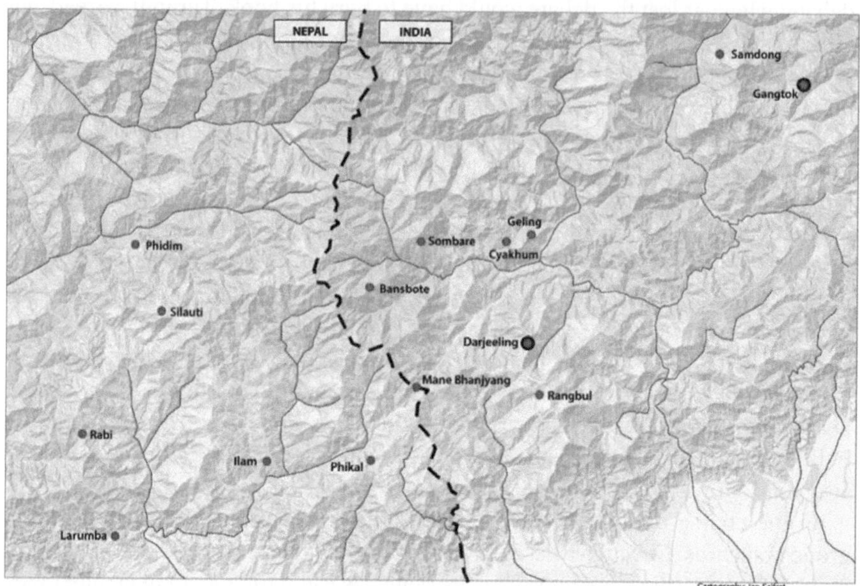

Fig. 5.1 Map of the Singalila borderland (Cartography: Jan Seifert).

contemporaries, the Josmani influence was strong there and also affected the next generation even after Gyan Dil's departure.

Phalgunanda, or Nardhoj Liṅgden (his birth name), was born in 1885 in Chukchinamba, a small Limbu village near the town of Rabi in Panchthar (for details of his biography, see Gurung & Dahal, 1990; Gaenszle, 2013). His mother died shortly after his birth, and he was a withdrawn child with early signs of spiritual leanings and possession (classical symptoms of a shaman). As a teenager, he left home in order to make some money. He first went to Gundri Bazaar, Darjeeling, where he could sell some local chicken. There, in British India, he eventually found work as a porter in the context of the Younghusband expedition to Lhasa (1903/1904). He carried loads up and across the Himalayan high mountains, walking on snow-covered trails, a risky task which some of his fellow porters did not survive.

Eventually, he left the village of his childhood in east Nepal for good and was recruited into the British Army in 1907 when he visited his brother, a soldier stationed in Burma. It is there, interestingly, that he became initiated into the Josmani order, 'on the banks of the Irawati', where a disciple of Shashidhar's line taught him Vedanta and gave him the name of Phalgunanda (Gurung & Dahal, 1990, p. 21). Josmani influence reached well into the eastern fringes of British India.

Military service provided new opportunities to travel abroad: in 1914/1915, Phalgunanda was sent to Europe and fought in WWI (probably in France) against the German Army, but he felt out of place. He saw the meaninglessness of war,

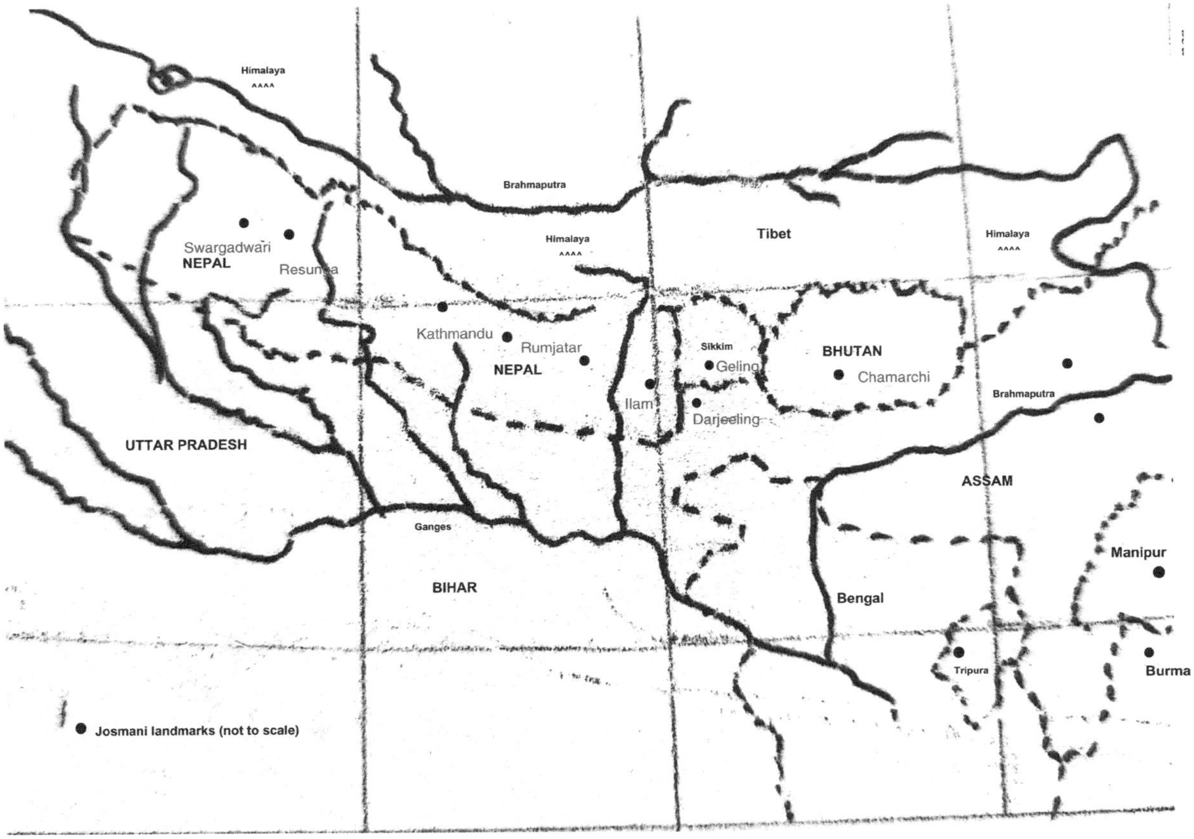

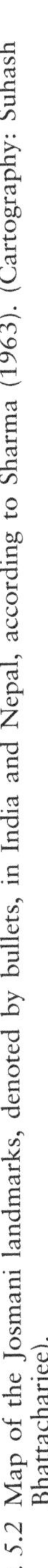

Fig. 5.2 Map of the Josmani landmarks, denoted by bullets, in India and Nepal, according to Sharma (1963). (Cartography: Suhash Bhattacharjee).

refused to shoot the enemy and left the Army. Eventually, he returned to India in 1915 and led the life of a *sādhu*, wandering through the subcontinent and the Himalayan mountains. His travels took him to the Northeast: Nagaland, Assam, and Bhutan (Gurung & Dahal, 1990, pp. 15, 17, 25). He also undertook a *Cār Dhām Yātrā* all through India: he visited Kashi, Mathura, Dvaraka, Rameshvar, Jagannath (Puri), Ayodhya, Hardwar – Badrinath, and eventually returned to Nepal, to his birthplace, Chukchinamba (Panchthar).

Thus, by the age of 35 years, Phalgunanda had seen quite a bit of the world! But once he finally returned to his home village, he had made up his mind and decided to focus his activities on his home region: Limbuwan. It is in the homeland of his ethnic community, the Limbu, that he eventually became not only a religious leader, but also a social and – especially as seen by the state – political activist.

Initially, he still undertook various travels to the Indian side of the border in the mid-1920s. In 1924, he undertook a trip to Darjeeling via the border pass of Manebhanjyang (Gurung & Dahal 1990, p. 36) with some disciples. He was already seen as a '*Mahātma*' – a person of charisma and fame, able to bring rain in the drought. This image brought him into conflict with the local 'Chitre Lama', a Tibetan Buddhist who challenged his authority. Another journey in 1926 took him to Darjeeling but also to Sikkim (Gurung & Dahal, 1990, p. 25). However, it seems that apart from these shorter trips, he had no more business in British India. His task was to reform the Limbu in Nepal.

Fig. 5.3 Image of Phalgunanda Lingden (Photograph by the author).

One important activity was temple building (something practically unknown in traditional Limbu culture). He built the first *manghim* ('god-house') in Charkhola Chukchinamba in 1929 (Gurung & Dahal, 1990, p. 23). Others soon followed: Panchthar Lalikharka (1931), Yangrup Kaveli (1935), Panchthar Silauti (1938), Athrai Nigradenken Chilingse (1939), and Charkhola Jitpur (1942). These are the temples founded by the Mahaguru himself; many others continue to follow his example to the present day.

The Limbu, at that time, were increasingly unhappy with the Shah monarchy under the Rana government. Due to the settlement of high-caste Brahmans and Chhetri, but also Magar and Gurung immigrants, there was increasing tension in the eastern hills: Limbu land under communal tenure was appropriated by these 'outsiders'. Many Limbu were indebted to moneylenders of high caste origin to whom land was mortgaged – and often they lost this land when the money plus high interest could not be paid back (Caplan, 1970).

There was strong pressure on the indigenous groups to assimilate into the dominant culture of Hinduism. Many of the Rai and Limbu felt that their religion was somehow lacking in prestige and needed to change, so they initiated reforms to counter the challenges of the fast-spreading Hindu culture. Phalgunanda, a charismatic and authoritative figure who had seen the world, became the leader of such a reform movement. His acclaim increased during the 1920s, and eventually, he organized an important gathering which made him popular even beyond the region.

In 1931, he convened a meeting of the Ten Limbu Chiefs (*das Limbu*) on Silauti Danda, a high ridge marking the boundary of Ilam and Panchthar districts – at a scenic hilltop overlooking much of Limbuwan. This meeting was probably the biggest political convention in east Nepal during the autocratic Rana period, and it created some nervousness in the capital at the time, as it signaled popular resistance. Above all, it had lasting consequences for the new reform movement and Limbu identity.

In his major address to the gathering, Phalgunanda exhorted the assembled Limbu to retain their ethnic identity but also to reform their indigenous religious practices. This meant, above all, the abstention from the consumption of alcoholic drinks (beer, brandy) and meat, which implied fundamental restrictions in ritual practice, as animal sacrifice and offerings of beer are an integral part of ancestral cults. Further, the reforms included the abolishment of the tradition called 'marriage by capture' (instead of arranged marriages), the prohibition of bridewealth payments, and a ban on dancing during funerals. Instead of incurring expenses on such 'needless' festivities, Phalgunanda asked his Limbu brethren to use their wealth to send their children to school so that they could learn to read and write. Importantly, he stressed that both girls and boys should be educated.

The result of this big meeting was a joint decision, known as the Convention of the Limbu's True Religion (*Yakthung Cumlung Satya Dharma Muculkā*), which committed the Limbu to follow Phalgunanda's reform proposals. This was the beginning of the Limbu's cultural reform movement, a self-conscious re-definition of Limbu identity, marked by a refusal to comply with dominant Hindu ideas.

But Phalgunanda's ideas also had an impact on the political process in the Kirat region in east Nepal. The reform of religion held out the promise of solving the wide-spread problems of indebtedness and land loss (see above). By adopting a new life-style, so it was claimed by the Mahaguru (as he was called), all the long-standing troubles could be extirpated: if there were no high expenses due to moderate living, there were also no debts. Thus, if the Limbu changed their religious customs and habits, they could retain, and even get back, their communal land which they had already lost to the high caste creditors.

In this way the reform movement was a political programme which significantly threatened the whole social order of inter-caste relations, hierarchies, and Brahman hegemony. Consequently, the high-caste elite – and the state – saw Phalgunanda as a dangerous seditionist, who was difficult to get hold of because he appeared in the garb of a religious person.

In fact, a major reason why he was seen with suspicion, and awe, was that he was constantly on the move, wandering from village to village accompanied by his disciples. He stayed in the lower hills during the cold season, staying at one of his temples and as a guest of his followers. In the hot season he moved to the higher mountains, doing pilgrimages by visiting remote sacred places. For example, in 1938 he undertook a big pilgrimage (*jātrā*) to Khumbakarna (Kanchenjanga), accompanied by a large number of devotees.

In the same year, after Phalgunanda had just returned from his Himalayan tour, some opponents accused him of sedition. A certain Kancan Giri in Dhankuta accused him of conspiracy against the state and treason (*rājdrohi*), because – so the accusation went – he wandered through the Limbu region with the purpose of instigating the people against His Majesty's Government and demanding a separate state. He was brought to court in Dhankuta and was questioned by the *baḍāhākim* of the District, Madhav Shumsher Rana. But the local people regarded Phalgunanda as a religious leader, a saint, and not as a political activist, and so, given the strong popular support, he was eventually released. But, Phalgunanda also had staunch enemies in powerful places. It was reported that he was taken to court in Ilam because he was said to have forced people to provide donations for his temples. This accusation, however, was later dismissed (Gurung & Dahal, 1990, p. 45).

Another innovation heralded by Phalgunanda in the later part of his religious career contributed to his fame among the Limbu – and simultaneously aroused greater suspicion amongst the authorities: during his final years he propagated the use of the distinct Limbu script, a set of letters which had earlier been invented by a Limbu and Sikkimese monk, Srijanga Singthebein, in the eighteenth century (see Vandenhelsken & Gaenszle, 2019).

In the early 1940s, Phalgunanda, it is told, discovered some mantric letters in a cave, and following this, he decided to study this forgotten script and disseminate it for the use of primary education (mainly promoting literacy) in the Limbu mother tongue. But this discovery did not come from nowhere. Phalgunanda must have heard about I.S. Chemjong, who had been propagating the same script in Darjeeling since the 1920s. From that point on, the Limbu script became the vehicle of a broader ethnic movement in both India and Nepal. We do not know much about the interrelationship between the two Limbu activists, who most likely knew about each other, though it is not clear whether they ever met. In any case, both acted in a complementary fashion: Phalgunanda focused entirely on the Limbu area in Nepal, whereas Chemjong stayed largely in Darjeeling, addressing the Limbu on the Indian side. Only much later, in the 1960s, did Chemjong eventually become a professor at Tribhuvan University in Kathmandu.

Phalgunanda's engagement in such a movement, which promoted the conscious use of a minority language and script, was a risky matter. It must be emphasized that during the Rana's autocratic rule, especially in the 1930–1940s, publishing books without government consent was equivalent to sedition and conspiracy. For fear of resistance movements, the Ranas did not allow private publishing; printing was restricted to state institutions only. Furthermore, the distribution of books, or even manuscripts, in an ethnic script was considered even more suspicious. In such a situation, cooperation with like-minded political activists in British India was a highly dangerous matter, and so it seems that in order to avoid further trouble with the state authorities, the Mahaguru refrained from transnational ties and stayed on the Nepal side of the border for the rest of his life.

To conclude this section, one might say that there was a kind of 'nationalist turn' in Phalgunanda's life course, sometime in the 1920s. Before that time, Phalgunanda led a rather cosmopolitan life, moving across the British Empire and wandering through the sacred geography of the South Asian subcontinent, Bharatvarsha, with little attachment to particular places. He was a true peripatetic, travelling from place to place in an imaginary religious landscape (Eck, 1998). But once he focused on the fortune of his brethren in Nepal (Limbus), he became more politically engaged and thus had to define his standing vis-à-vis the state (*sarkār*) and Nepali society in general. He was not against the state, but wanted acknowledgement of Limbu cultural achievements within the kingdom. Whether self-consciously or not, he became a fighter for a new (ethnically inclusive) Nepali nationalism. This is the reason why in today's Nepal, he is regarded – and celebrated – as a 'national luminary' (Gaenszle, 2021).

Conclusion

The two cases of itinerant religious leaders can be taken as examples of the unique powers deriving from the vagrant lifestyle of such gurus. I should mention that there were other bhakti-inspired, anti-brahmanic sects which were successful among the Kirat groups in east Nepal during the reign of the Ranas (and before): there were Kabirpanthis (e.g., in Chintang), followers of Yogamaya (Arun Valley; see Aziz, 2001; Hutt, 2013), and Vaishnavas (e.g., in Varahaksetra). But few of their leaders had such a long-standing impact.

The cases presented here illustrate two aspects of this kind of religious itinerancy. First, they show that religious community-building worked quite well across the Singalila borderland well into the twentieth century. The political boundaries separating Nepal and British India were constrictive but far from being impervious; they could be used for one's benefit in case an evasion of state control was necessary. Ideologically, the boundaries were not of great significance for the gurus, as the addressees of their proselytisation were people living on both sides of the border. Second, it becomes clear that the mobility of the leaders was a big advantage in dealing with the state. Though both leaders were not directly acting against the rulers and the state, and even had, at times, amicable relationships, their link with statehood was conflict-fraught and tense.

In any case, one has to be aware that the situation changed over time, and the developments have to be put in historical context. Initially, especially in their youth, both itinerant leaders were mobile to a considerable degree, exploring unknown places, thus gaining a broad perspective on the social and political situation of their communities. At a time without mass media communication, it seems these leaders were important conduits of news about the outside world. They brought in new ideas of social reform, values of equality, education, and morality. As far as their ideology is concerned, they cannot be seen as supporters of nationalism but had a more transnational outlook. The Josmani, to be sure, were not bound to Nepal, though the Nepali language played an important role in their activities; the Satyahangma followers, though not directly anti-state, had a problematic relationship with the Nepal government, which regarded them as a 'seditious' resistance movement, so they looked for allies outside.

However, the situation clearly changed in the course of the twentieth century, and national boundaries increasingly impeded the earlier spatial mobility. For the Josmani, this meant that the vital communities outside Nepal lost contact with those in the Hindu Kingdom, leading to an erosion of tradition among the latter. In the case of the Satyahangma community, the strong attachment to Limbu and more generally Kirat identity and the struggle for minority rights got them increasingly (especially since 1990) involved in Nepal's ethnic politics. Also, on the Indian side of the border, ethnicity increasingly became a crucial issue, but the general social and political context in Darjeeling and Sikkim was quite different from Nepal (on Sikkim cf. Vandenhelsken & Khamdak, 2021).

Finally, let us take a look at the situation today. As already mentioned, there are only a few Josmani left in Nepal, mainly in the region of Rabi (Panchthar). The Josmani as a community are now mainly existent in India, especially in Sikkim. There is a lively community in the village of Geling (West Sikkim), the place of Gyan Dil Das's *samādhi*. The temple area has been given generous support by the earlier Chief Minister of Sikkim, who was in office from 1994–2019, Pawan Chamling (Rai). These favours were partly due to his close links to the area of southwest Sikkim, where his forebears come from, and many of the Rai reside. Some of the Rai are also among the Josmani followers.

As to the Satyahangma, there are only a few communities outside of Nepal, mainly in Darjeeling (Bansbote, Kurseong), as well as in Sombare, Sikkim. The main centre of their community is the 'Dharmaguru' Atmananda in Larumba (Ilam District), who is busy building a 'Spiritual Centre of Kirat' – a kind of 'Kirat Vatican', as one follower put it. I met some followers of his in Darjeeling who regularly cross into Nepal to visit him in Larumba. But clearly, the major support comes from the large Kirat overseas diaspora: from the UK, USA, Hong Kong, and Singapore. In spite of the large Nepali diaspora in India, there seem to be few followers in other parts of India. Symptomatic of the current active expansion of the Satyahangma religion in Nepal is a real estate development project near Damak (Jhapa District) called the 'Satyahangma Colony': it invites investors from around the world – but, surprisingly, not from India.

Thus, what we see in these more recent developments is an erosion of religious itinerancy and a tendency toward a sedentary lifestyle: today's followers of the earlier gurus (including their successors) have settled down and built temples and monasteries. This tendency was already visible in the policies of the Rana state, which clearly attempted to control the potentially 'dangerous' movements of local itinerants. Thus, it was not so much a colonial administration or mindset which restricted itinerancy and enforced settlement but the process of modern state-building guided by the ideals of effective government and the generation of revenue (taxation). This formation of the modern nation-state (cf. Burghart, 1984) goes hand in hand with a kind of nationalisation of religion, which can be seen as linked to the general nationalisation of space. As borders have become more impervious due to stricter control of passage and regimes of citizenship, so religious affiliations depend on and create new boundaries. In spite of globalised communication practices, the religious leaders are now commemorated as national heroes and have thus become a national asset.

Acknowledgements

Parts of this research have been carried out in the context of the project 'Transborder Religion' (Austrian Science Fund, FWF, P29805), headed by Mélanie Vandendelsken. I am grateful to the funding agency, the team leader, and Prem Chhetri for their support. I also thank Bairagi Kainla and Novel Kishor Rai for their long-standing assistance.

Notes

1 In the context of Ramanandis, he writes: 'According to the Renouncers …their itinerant movement is a representation of the movement of the Invisible Spirit of the universe' (Burghart, 1983, p. 363).
2 On theoretical discussions of Nepal-India 'border(land) people' following the seminal article by Baud and van Schendel (1997), see Gellner (2013) and Shneiderman (2015).
3 Phalgunanda was declared the '16th National Luminary of Nepal' by the government's cabinet on 1 December 2009, and thus shares the status of a '*rāṣṭriya vibhūti*' with such figures as King Janak, Gautam Buddha, King Prithvi Narayan Shah, and the national poet Bhanubhakta Acharya.
4 I use the term 'religion' in the loose sense common in South Asia, i.e., not implying the strict institutional framework of a 'church', but rather the religious practice of an established and inherited custom, as expressed in the notion of *saṃpradāya* or *paramparā*. Though these words are sometimes rendered as 'sect', that term should be used with reservation as the Indian context does not imply a heterodoxy deviating from an orthodoxy.
5 In the latest Nepal Census of 2021, the followers of the 'Kirat' religion are given as 3.2% of the population.

References

Aziz, B. (2001). *Heir to a Silent Song: Two rebel women of Nepal*. Centre for Nepal and Asian Studies.

Baud, M., & Schendel, W. V. (1997). Toward a comparative history of borderlands. *Journal of World History*, *8*(2), 211–242.

Bayly, C. A. (1996). *Empire and information: Political intelligence and social communication in North India*. Cambridge University Press.

Burghart, R. (1983). Wandering ascetics and the Rāmānandī Sect. *History of Religions*, *22*(4), 361–380.

Burghart, R. (1984). The formation of the concept of nation-state in Nepal. *Journal of Asian Studies*, *44*(1), 101–125.

Caplan, L. (1970). *Land and social change in East Nepal: A study of Hindu Tribal relations*. Routledge and Kegan Paul.

Eck, D. (1998). The imagined landscape: Patterns in the construction of Hindu sacred geography. *Contributions to Indian Sociology*, *32*(2), 165–188.

Gaenszle, M. (2013). The power of script: Phalgunanda's role in the formation of Kiranti ethnicity. In V. Arora & N. Jayaram (Eds.), *Routeing democracy in the Himalayas: Experiments and experiences* (pp. 50–73). Routledge.

Gaenszle, M. (2021). Satyahangma rituals: Commemorating Phalgunanda in Eastern Nepal. *European Bulletin of Himalayan Research [Online]*, *57*.

Gellner, D. N. (Ed.). (2013). *Borderland lives in Northern South Asia*. Duke University Press.

Gurung, H. B., & Dahal, K. (1990). (2046 V.S.). *Tapasvī Phālgunāndako Jjīvanī*. Bhakti Prasad Tumbapo.

Gustavsson, L. (2013). *Religion and identity politics in the Indian Himalayas: Religious change and identity construction among the Limboos of Sikkim* (M.A. thesis).

Hutt, M. (2013). The disappearance and reappearance of Yogmaya: Recovering a Nepali revolutionary icon. *Contemporary South Asia*, *21*(4), 382–397.

Lecomte-Tilouine, M. (2002). The history of the Messianic and Rebel King Lakhan Thapa: Utopia and ideology among the Magar. In D. N. Gellner (Ed.), *Resistance and the state: Nepalese experiences* (pp. 244–277). Social Science Press.

Pinch, W. R. (2006). *Warrior ascetics and Indian Empires*. Cambridge University Press.

Rai, N. K., & Dumirai, N. (2013). (v.s. 2069). *Josmanī Dharma Sampradāya ra Darśan*. Nepāl Prajñyā Pratiṣṭhān (Nepal Academy).

Sharma, J. (1963). (2020 V.S.). *Josamanī Santa-Paramparā ra Sāhitya*. Royal Nepal Academy.

Sharma, J. (2017). Kalimpong as a transcultural missionary contact zone. In M. Viehbeck (Ed.), *Transcultural encounters in the Himalayan borderlands: Kalimpong as a "contact zone"* (pp. 25–53). Heidelberg University Publishing.

Shneiderman, S. (2015). *Rituals of ethnicity: Thangmi identities between Nepal and India*. University of Pennsylvania Press.

Timalsina, S. (2010). Songs of transformation: Vernacular Josmanī literature and the Yoga of cosmic awareness. *International Journal of Hindu Studies*, *14*(2–3), 201–228.

Vandenhelsken, M. (2021). 'Ils ont transformé la divinité Yuma en Dieu!' Recompositions Religieuses et Dissensions chez les Limbu du Sikkim. *Ateliers d'anthropologie - Open Edition Journals*, *49*, 1–15.

Vandenhelsken, M., & Gaenszle, M. (2019). Limbu: Limbu religion and the remaking of the community. In M. Carrin, M. Boivin, G. Toffin, P. Hockings, R. Rousseleau, T. Subba, & H. Tambs-Lyche (Eds.), *Brill's encyclopedia of religions of Indigenous people of South Asia*. Brill (Online version).

Vandenhelsken, M., & Khamdak, B. L. (2021). Loyalty, resistance, subalternity: AHistory of Limbu 'participation' in Sikkim. *Asian Ethnicity*, *22*(2), 235–253.

White, D. G. (2009). *Sinister Yogis*. Chicago University Press.

Yākthumbā, M. (2018). *Ko hun Kirāt ra Kirātlimbū? Ke ho Yumādharma?* Bimalā Niṅlekhu Yākthumbā.

The Pre-Colonial and the Scriptural

6 'Floating Straight Obedient to the Stream'

Bibliomigrancy and Riverine Journeys in Colonial Bengal

Swati Chattopadhyay

British colonial travel narratives of India rested on certain rhetorical conventions: *arrival and first view* at landfall, travel through the interior of a foreign land while the *surroundings are revealed as landscape,* and the viewer standing on an *elevated location* from which to survey the land. Here the 'traveling gaze', to use David Arnold's phrase, conflated the tasks of military reconnaissance, the aesthetic pleasure of the picturesque and sublime, and the administrator's view of land as a source of revenue awaiting improvement (Arnold, 2014).

The riverine regions of the lower Gangetic plains posed difficulties to such colonial representational strategies. The practical difficulty of navigating tortuous streams of the delta that frequently changed course and the 'desolate' landscape on arriving at the coastal reaches of the Bay of Bengal were imperfectly suited to the prevalent aesthetic conventions of the picturesque and sublime. Refracted through the lens of administrative imperatives and military reconnaissance, such land and river forms instead came to be perceived as obstacles to colonization – dangerous and unmappable (Chattopadhyay, 2005, pp. 89–109).[1] In effect, in colonial eyes, these regions became evacuated of the meaning of landscape as sensible occupation, that is, the process through which land is made into landscape. European visitors arriving near the coastline of the Bay of Bengal routinely described it in terms of an absence of landscape (Chattopadhyay, 2005, pp. 26–28).

In this chapter, I pursue two strands by which vernacular aesthetic conventions were inaugurated to confer value on this regional topography. These aesthetic conventions do not so much turn upside down the conventions of the picturesque and sublime, as they convey a different set of preoccupations with the materiality of land and water by crossing languages and creating new vantages. While my investigation is exploratory, at one level, the task of this chapter is circumscribed. I plan to examine the description of aqueous journeys in two Bangla-language texts – a novella by Bankimchandra Chattopadhyay (1838–1894) titled *Kapalkundala,* published in 1866, and a memoir of childhood days, *Amar Ma'r Baper Bari*, published in 1972, by Rani Chanda (1912–1997) – to consider the surprising reveals and roles of water travel in Bengali literature as they memorialize the landscape of a colonial era. In a broader sense, the juxtaposition of a novella and a memoir is intended to invoke two different strategies of world-building in which bibliomigrancy, that is, movement between languages, is correlated with movement through space to create a country imagination. Written a century apart, each invokes a past landscape

DOI: 10.4324/9781003544524-10

to draw from it the elements of world-building in ways that challenge colonial ways of seeing the land as landscape. *Kapalkundala* is set in the early seventeenth century before British colonialism had gained a foothold in the region and thus returns to the precolonial world to craft the landscape as a sublime (proto-nationalist) revelation. *Amar Ma'r Baper Bari* travels back to the second decade of the twentieth century to a lost landscape of childhood days of the author as a tribute to her mother's memory, and in so doing inaugurates a landscape temporality that is attentive to the ephemeral.[2] Such a mode of reckoning with time bypasses both colonialist and nationalist sublimes.

Surely, the two texts conveniently signpost the century in which Bengali literature came into its own as a modern endeavour. Given that there is a considerable body of Bengali literature that uses river travel as a way of telling stories, autobiographical and fictional, my choice of these texts is motivated by what I perceive as their risky engagement with the utterly mundane in thinking about the land/water relationship and in their response to the past as a landscape.[3] Both texts inaugurate new descriptions of the land. Each story is marked by a caesura that forces them to 'float obedient to the stream', and thus take their respective narrative chances. In *Kapalkundala* it is literally a lost ship, and in *Amar Ma'r Baper Bari* it is the death of the mother (and implicitly a motherland). I focus on the description of coastlines, river passages and waterscapes of lower Bengal, and the keenness with which these texts find in the riverine geography the opportunities to recover a lost landscape. Aqueous passages play an important role in the narratives, despite the fact that neither of these are travelogues.

Bibliomigrancy

I borrow the phrase 'bibliomigrancy' from B. Venkat Mani's discussion of the emergence of German *Weltliteratur* in the early nineteenth century when a new world could be envisioned through colonial exchanges and a new idea of world-in-print could be accessed from Europe. 'Accessibility to the world was not merely a function of travel', he notes, 'the world was fast becoming accessible *in print*' (Mani, 2017, p. 66). Long before the mass migration of populations from Asia and Africa to Europe, the migration of ideas from these continents in print and manuscript form took place at an unprecedented scale, fuelling the norms of comparative discourse. The eager interest with which original manuscripts in Arabic, Chinese, Persian, and Sanskrit works and their translations into German and English were sought out by German scholars is the crux of Mani's argument (51–52). In this movement of books and migration between languages, we may identify the beginnings of a number of comparative fields of study – comparative religion, literature, and history – and indeed the implicit or explicit comparative modality through which modern humanities and social science disciplines were shaped. Mani notes that in the absence of a nation-state in Germany, this sense of 'worldliness' in the German-speaking world was in competition with a sense of 'national' community and was shaped largely in terms of language and literature (67).

One could argue that in the second half of the nineteenth century, a similar process was taking place in Bengal, albeit from a decidedly marginalized position of subalternity to Europe and European literature. When in the mid-1830s Thomas Babington Macaulay's policy defunded education in the Indian classical languages in favour of English, as Mani notes, he not only expelled 'literary traditions in Sanskrit and Arabic from the world literary landscape', but he also delegitimized the myriad other indigenous languages and literatures that populated the British colonial territories (61). It is this literary field marked by the epistemic violence of colonialism that Indian authors and intellectuals inherited and set about to fix and resist by reshaping native tongues into new literary fields.

Bengali authors cultivated a sense of worldliness through borrowings and translations of works from other languages while actively recuperating what they considered to be 'lost' literary practices. This included early modern Bangla poetry and folklore. No stranger to polyglossia or mixed literary forms in which multiple languages were incorporated, the Bengali literati expanded their knowledge of Persian, Sanskrit, Hindustani, and Bangla to European languages, first for purposes of commerce and administration, and second to cultivate intellectual ambitions.[4] The demand for access to works from European languages would promote a bibliomania resulting in a phenomenal growth of print culture and libraries in Bengal.[5] For Bankimchandra (hereafter Bankim) in the days of nineteenth-century bibliomigrancy and for Rani Chanda, who was a beneficiary of the print culture that flourished in Bengal in the first decades of the twentieth century, movement between languages was expected. They deploy literary migrancy toward different ends and follow distinctly different contours. However, their writings, precisely because they are separated by almost a century, catch not only two strategies of bibliomigrancy in which travel as metaphor and process illuminate movement between language worlds, but they also signal two imperatives of decolonization.

More specifically, Bankim, who was highly critical of contemporary Bangla literary production – most of which he did not deem to contain any literary merit – chose to juxtapose European-language worlds with the landscape imagination of classical Sanskrit, Hindustani, Maithili, and Bangla, as well as the spoken Bengali vernacular, to inaugurate a modern Bengali literary imagination.[6] Much has been written about Bankim's anticolonial nationalism, before nationalism became a creed. He was deliberate in breaking established norms and creating new ones. It is scarcely noticed that in so doing Bankim also created what I would call the first pass at viewing the land through a modern vernacular literary landscape aesthetics. The vernacular itself is recast at this time to do the work of decolonizing literary imagination, one that both makes references to but is not beholden to Western literary norms. This landscape aesthetics is produced through the incessant movement of his characters from one place to another – their passage allows him to describe the land, whether these are majestic forested terrains or everyday rural life of the Bengali landlords and peasants. Bankim's novels have a pronounced sense of spatiality precisely because they are about movement between different worlds and deliberate about registering the surroundings and atmospheres as he narrates different stages of travel.

Bankim joined the provincial service of the colonial government in 1858 as deputy magistrate and collector and retired in 1891. His employment meant being 'posted' in *sadar* (district headquarter) towns in provincial Bengal, and the job required him to travel for inspection visits.[7] It is through such travel that he became familiar with the geography of the Bengal Presidency beyond his ancestral village, Kanthalpara, in Hooghly, and Calcutta, the city where he was educated and where he lived after retirement.[8] His memorialists note his familiarity with the rivers and canals of the delta as well as the seacoast of Medinipur.[9]

His novels feature both travel across space and across time. In terms of geography his novels are populated with elaborate descriptions of jungles and rivers as settings (as in *Durgeshnandini, Kapalkundala, Anandamath*, and *Debi Chaudhurani*) and his penchant for historical references meant that the stories move far afield from lower Bengal to the heartland of Mughal-Pathan-Rajput rule in northern and western India. Bengal as a Mughal province, and lower Bengal and Orissa as its farthest reach, offers a narrative-geographical space with differentiated valences. The characters move from locales of power – Delhi Agra, Murshidabad – to marginal geographies in the riverine peripheries where the stories culminate. *Kapalkundala*, only his second novel, brings together distinct land and water geographies: sea, river, coastal scrubland, jungle, and village to speculate on what becomes of life when untempered by social and political intrigues and domestic restraints. There is a certain interchangeability between the rugged beauty of dunes and shrubland and the character of Kapalkundala, whereas the male protagonist in the novella is portrayed as one who has lost his bearings in this landscape, removed as it is from the settled sphere of village life. The coastline is an exotic and liminal space between the open ocean and the village.

In contrast, Rani Chanda (hereafter Rani) recalls yearly seasonal journeys from the city of Dhaka to a remote village via the waterways as a way to tell the story of her mother and her mother's natal home – a sort of return to an originary home space, when such returns were no longer possible. As she travels with the water, she moves between the standard literary form of Bangla and the colloquial Bangla of the village that localizes and marks deviations from the norms of intelligibility. This allows her to access the materiality of the spaces – the textures, sounds, and pace of ordinary life – in which to locate the figure of her mother. The aqueous geography is itself mobile and lends a dynamism to the spaces she describes. It is both routine and full of surprises. As her story pauses on the small things of everyday life, we are introduced to the life of a Bengal village as if for the first time.

Rani's original inspiration to write about the small village had come from Rabindranath Tagore. A young Rani spoke of her mother's village with such enthusiasm that Rabindranath asked her to write down these remembrances. But she didn't. 'What is there to write about?' she had thought. The stories, sparkling with childhood memories though pleasant in conversation, seemed commonplace, unimportant, ill-suited to a life in ink and print. Only after her mother passed away did she start committing her memories to paper. Rani's imperative, however, is more than personal. Her invocation of the rural dialect of eastern Bengal is shadowed by loss: of the house to which she would never return, but also by the

Partition of 1947 that split Bengal along Hindu and Muslim lines and in so doing alienated home for so many. For someone who had taken an active part in the nationalist movement, the promise of nationalism as freedom from colonialism remained just that – a promise, now burdened with a grievous loss. Decolonization after Indian independence from British rule remained a continual process of critical self-reflection.

Both texts are concerned with home and domesticity as imaginary grounds. In *Kapalkundala* it appears as the limits of domestic life by first travelling away from home and then in a (futile) attempt to return to that home space. In *Amar Ma'r Baper Bari* home figures as the reconstruction of an already lost space, narrated through loving details of the day-to-day goings-on in a small village.

Journey from Home

Kapalkundala, set in the early seventeenth century, invokes the moment of shipwreck in William Shakespeare's *Comedy of Errors* as an epigraph to launch a narrative of pilgrimage, abandonment, discovery, and loss, taking the Bay of Bengal and its coastline as a frame. The novella is continually signposted with references to English, Sanskrit, and Bangla literature to create a modern landscape sensibility anchored in historical references. This is how Bankim begins the story:

> About 250 years ago on the early hours on a day of the month of Magh a passenger boat was returning from Gangasagar. It was customary for such boats to travel in groups in fear of the Portuguese and other pirates. But this boat was traveling by itself. Because a deep fog had engulfed the horizon in the last hours of the night. Not being able to orient themselves, the boatsmen had strayed far from the rest of the group. At this time, no one knew in which direction the boat was headed. Most of the passengers were asleep. Only two men—one elderly and one young—were awake.
>
> (Chattopadhyay, 1953, p. 73)

While the elderly gentleman, in fear of losing his life, becomes impatient with the boatsmen, blaming them for the imminent disaster, the young man, Nabakumar, seems unperturbed. Nabakumar tells his elderly companion that if he is so worried about the uncertainties of travel, he shouldn't have taken this trip: one could just as well earn religious merit from home as one could by going on a pilgrimage. The elderly man is eager to return home because he had lately received news that brigands (referring to the Pindari raids under Maratha aegis) had robbed his winter harvest. The dueling concerns about the protection of house and property and the desire to accumulate religious merit by undertaking a pilgrimage inflect the elderly man's view of the danger at sea. He in turn asks Nabakumar, 'then why did you come on this pilgrimage?' Nabakumar responds:

> I told you already, I had a desire to see the ocean. Afterwards he spoke in a soft tone: Ah what a sight, I won't forget it in this life and the next:

Duradayashakranivashya tanwi
tamalatalibanaraji neela.
Abhati bela labanamburaashe
dhwaranibadheba kalankarekha.

(73)

[The distant seashore stretches in an arc slender/a blue-black line of tamal and palm forest/. The shore of the salty mass of water appears as a dark-stained rim.]

The verse is from Kalidasa's epic Sanskrit poem, *Raghuvamsa* (ca. 5th c. CE), in which Rama and Sita are gazing at the coastline of peninsular India. While it is tempting to impute a proto-nationalist geo-sensibility in Bankim's use of this quote to describe the shoreline of the Bay of Bengal, I wish to resist that temptation to discover something more important in terms of what travel means in this novel. For Nabakumar this is clearly a secular pilgrimage of a different sort. He is an aesthete and a romantic. The aesthetics of travel is articulated through recourse to a classical text. But Bankim's world-building is framed by texts from various times, spaces, and languages: each chapter contains an epigraph, and apart from Shakespeare and Kalidasa, these are drawn from the works of the poet Vidyapati (fourteenth-fifteenth centuries, who wrote in Maithili and Sanskrit), the eighteenth- and nineteenth-century Romantic English poets John Keats, Lord Byron, William Wordsworth, Thomas Babington Macaulay, and closer to home, his nineteenth-century contemporaries: the Bengali poet, Michael Madhusudan Dutta (who wrote in both Bangla and English), and the playwright Dinabandhu Mitra who wrote in Bangla. The reader is enjoined to expand their literary bounds and explore the landscape affected by references to different literary and aesthetic worlds, in which historical fiction, romance, and epic poetry intersect. Bibliomigrancy helps create a frame – a scaffold within which to render the desolate seashore a worthy locus of high literature. The first set of chapters of the story is set with subtitles such as *sagarsangam'e* (at the confluence of the sea), *upakul'e* (on the coast), *bijon'e* (in solitude), *stupashikhar'e* (on top of the dunes), and *samudratat'e* (on the sea beach) to mark the travel space.

What follows next in the story is that at daybreak, with the fog burning off, the boatsmen locate the coastline and decide to cast anchor and wait for the tide. Nabakumar goes on to the shore to fetch firewood. A sudden tide impels the boatsmen to lift anchor and sail toward home. Nabakumar gets forsaken on the desolate shore. Bankim's description of the shore this time is very different. It is no longer a shoreline that seemed beautiful from a distance:

At that time there was no sign of human habitation here. It was forest land, but of a peculiar type—from the mouth of the Rasulpur River and the Subornorekha it is a few *jojana* stretch of sandy dunes. Any higher, we could have called these hills. People here now call them *baliari*.

(76)

Naming landforms and calling out the specificities of the vegetation make the land legible, even if dangerous:

> From a distance the white summits of these dunes under the mid-day sun give off an unusual glow. On the lower reaches of these dunes are low forests … in which *jhati, ban jhau* and wild flowers are prevalent. Nabakumar observed that … there are no villages, no humans, no food, nothing to drink, the river's water is exceedingly salty…. No shelter from the icy cold wind. At night the danger from tigers and bears is real. Death seems certain.
>
> (76)

On the second day Nabakumar encounters a *kapalik* – an ascetic who gives him shelter but also plans to ritually sacrifice him to fulfill his obligations to the goddess Bhavani. The following day, when Nabakumar decides to flee from the ascetic and find his way home, he walks across the dunes and enters the jungle. Unable to orient himself in the jungle, he is revived by the sound of sea breakers. The sea is indeed nearby. He arrives on the beach and sits down. Here we see Nabakumar's eyes moving over and along the coast, the 'surfing, blue, endless ocean':

> On either side as far as the eyes could travel stretched a line of foam formed by the crashing waves. Like a garland strung from a heap of blemishless blossoms this white foam line has been laid down on the golden coast. In the far distance a European merchant ship lay like a large bird on the ocean's breast.
>
> (78)

It is difficult to say, Bankim writes, how long Nabakumar sat there watching the ocean. Here the sense of duration is severed to create a moment of singular importance. As evening gathers, he decides to return to the ascetic's ashram. On turning, he sees the awe-inspiring figure of a woman. In the dim light of the evening, she appears 'as if a painting on a canvas'. (78) A one-paragraph description of her beauty follows, particularly her long black loose hair draped over her shoulders. Nabakumar is dumbstruck by her sudden appearance and beauty. After a long pause the woman asks, 'traveler, have you lost your way?' (79)

The female figure is Kapalkundala, raised far from society on this desolate shore by the ascetic. Nabakumar's encounter with Kapalkundala creates the liminal moment of recognition and sets off a chain of events where travel becomes tropic: Kapalkundala saves Nabakumar from the ascetic; they escape from these shores to the town of Saptagram; Lutfunnisa (who happens to be Nabakumar's former wife whose father converted to Islam) moves from Agra to Saptagram; there is a constant movement between the town and the adjoining forest where Lutfunnisa and Kapalkundala, Nabakumar, and the ascetic meet. The narrative stretches from the northern plains to the Bay of Bengal and rides on descriptions of journeys, symbolic of the individual's journey to self-realization and his/her path through the thickets of societal norms and ethical dilemmas. Generous doses of historical details locate the story in a specific time and place to create narrative depth. Paths here, however, are not just carved out by the characters' movements in and through historical byways; paths get accidentally crossed to create narrative twists and turns.

All this does not end well. Kapalkundala and Nabakumar return to the coastal reaches where the Ganges meets the sea, and Kapalkundala standing at the edge between land and river, is swept away as the sandbanks give way to the crashing waves. Nabakumar follows (115). While the reader is brought back to the desolate shore, indeed the coastal space used as a cremation ground, in a completion of the narrative arc, neither of these two characters in the end return home. They are destined to be unhomed: being forced to negotiate their place in society, they lose. Here refuge and abandonment switch meaning as the shore from which Nabakumar was rescued becomes his moment of self-realization on the brink of death. The distant view – a clearly delineated line marking the boundary between land and water – that prompted the moment of aesthetic appreciation in the beginning is reversed in the end as an embodied sense of loss and submersion where the boundaries between land and water become indistinct.

The eponymous Kapalkundala, who delivers Nabakumar from certain death when he is forsaken, and who ultimately is forced into self-sacrifice, though the raison d'être of the story, remains a liminal figure. She personifies natural goodness and beauty, exemplified by her willingness to help and provide care, moving against all societal norms. This propensity to provide succour is innate, spontaneous, and cannot be curbed. In the end, because this propensity does not figure well within a society that teaches selfishness and fearful adherence to social mores, she must herself abandon her temporary home, family, and place. The story and its journeys are figured around her, but she does not fit. In this she embodies the landscape of this coastal region itself.

The loose tresses of Kapalkundala that define her and that Nabakumar notices at their first encounter become a signifier of Kapalkundala's unworldliness as well as her uncompromising independence. When cautioned that her venturing out at night might make Nabakumar unhappy, she remarks: 'If I knew marriage for women is slavery, then I would never have married'. (104) She cannot fathom why her movement has to be restricted for someone's happiness or peace of mind. The only time she tries to hide something from Nabakumar – a written note tucked into her coiffed hair – the tresses betray her. The note falls to the ground where Nabakumar discovers it and takes it to be a sign of intrigue. It becomes the starting point of his journey of distrust that ends in tragedy. (109)

The coastal land lies not only in between settlement and the open sea but also acts as a refuge for those who do not fit in society or cannot belong. Kapalkundala, incapable of intrigue, cannot be assimilated within the social. She appears only as a vision of what is possible but cannot be grasped. The reader is taken on a journey through selfishness, cruelty, danger, abandonment, societal oppression, corrupt governance, insipid domesticity, and surprising beauty to be shown a glimpse, merely a possibility, of goodness, faith, and freedom.

Jangal

While Bankim was stationed in Negua subdivision (*mahakuma*), present-day Kanthi, he lived in a bungalow on the coast. There he encountered a kapalik who he believed lived in the coastal jungles nearby. He seemingly chose the coast and jungle as locations for unfolding his story about natural-societal bonds and character formation because he found in this coastal country an unconventional beauty that did not accord with prevailing landscape aesthetics (30). Lacking a literary language to express its eerie, forlorn beauty, he thus returns to a precolonial moment to make a fresh start at description. This journey to the past invokes a land that shows signs of an empire embroiled in factional intrigue. The 'jangal' as a space of refuge and norm-breaking activities in his novels represents the unsettled moment of the precolonial empire.

Jangal, as a land designation and trope, has a complicated history in Bengal. In precolonial Bengal, the word *jangal,* from which the English word jungle is derived, had a range of meanings from grass savannah and dry scrubland, to land that had been left uncultivated and was therefore unkempt (Dove, 2003, pp. 107–109). This was arid land but fit for human habitation and growing crops, and appeared in surveys and legal codes on land tenure. Nathaniel Brassey Halhed, in his *The Code of Gentoo Law* (1776), refers to the extant legal status of jungle as land that has been left uncultivated for five years or more. Halhed translates jungle as 'waste'. By the late eighteenth century, the idea of clearing and enclosing wasteland in England had been brought within the moral imperative of improvement that justified enclosure and appropriation of common land from the peasantry.[10] The translation of an idea of waste from England to India gathered other connotations not evident in the English use of the term. Notably the idea gained a new importance in its application in the colonies to justify imperial expansion.[11]

In a late-eighteenth century military report on the Jungle Terry (Jangal Terai), we see the term jungle being used for land that remains uncultivated or has been abandoned, and for dense bamboo growth as well as thick grassland. Seen through the lens of warfare, the jungle is an obstacle, impeding passage and visibility. Here it is associated with unhealthiness (Browne, 1788, pp. 11, 25, 54, 66). Lord Cornwallis in his advocacy of the Permanent Settlement of Bengal, used the term jungle to refer to 'land inhabited by wild beasts from which no rent could be extracted' (Hoop & Aurora, 2017, p. 87). Beyond the failed opportunity to yield revenue, he viewed such tracts of unreclaimed land as evidence of native laziness – unwillingness to labour – and their their moral laxity. Jungles demonstrated the native incapacity to govern themselves (Gidwani, 1992, pp. PE 39–PE 46). In the Sunderbans, the visual and physical impenetrability of the jungle came to define the lower Gangetic delta. In the 1760s, James Rennell used both 'jungle' and 'woods' to describe the geography of the Sunderbans, whereas W.W. Morrieson in the second decade of the nineteenth century used the term 'jungle'.[12]

Thus, by the nineteenth century, the term 'jungle' was being used in English in colonial government reports as a catchall term for scrubland, dense forests, and hunting ground, and often used interchangeably with 'woods' and 'forests'. While in Hindustani and Bangla usage all these meanings were implied, there was a wider set of terms used to designate what came to be called wasteland, encompassing a range of land types. Importantly, the rules of designating wasteland changed over the course of the nineteenth century. With the establishment of the Imperial Forest Department in 1864 and subsequent laws that extended state sovereignty over forest lands, the forest emerged 'as a specific category of wooded land or land reserved for the production of woody crops'. In contrast, 'jungle' came to stand in for 'residual wooded land (which may be largely trees and shrubs considered less valuable)' and 'also unkempt, poorly managed, dangerous, weedy and in many other ways devalued or deprecated land' (Shivaramakrishnan, 2022). That is, despite the revenue-generating capacity of jungles or the exercise of usufruct rights in them (to gather honey, wood, game, and for swidden cultivation), in both the official language of administrative records as well as in popular use such as memoirs and gardening guides, the jungle slid into a figure of not simply uncultivated/unkempt land but a threat to the spaces of civilization.[13] As the use of the term was extended from India to other spaces across the empire, jungle came to embody 'the outer extremes of colonial expansion', even when such spaces were physically proximate (Weisberg, p. 173). The jungle was deemed to lack 'utility' from a very specific modern state-centric point of view, and it is in this respect that the jungle came to be seen as interchangeable with the term 'waste'. We can observe this switch in William Hunter's characterization of the 'jungle' in Cornwallis's text as 'waste'. Attribution of the category of 'waste' to the land implied its reclaimability by the state, taking it out of reach of inhabitants who used such land for subsistence or profit.

Bankim would have been well-versed in all these uses of the term. In his novels, jungles appear as a peripheral space for sure, but it is in this peripheral space, either forested land or erstwhile habitation engulfed in vegetation, that he locates his romantic sites of rebirth and rebellion. That the peasantry fled to the jungles to seek protection against oppressive landlords was a familiar story to him. It is in this remaindered space – *baaze jamin* in Persian/Bangla, literally meaning badlands – where his characters in *Kapalkundala, Durgeshnandini, Anandamath, and Debi Chowdhurani,* meet by design or accident and then re-form themselves into agential entities. His stories need to pass through such spaces to achieve their narrative ends. Jungles are for Bankim transformative spaces. There is an attendant function of these jungle settings: they set the temporality of the stories.

Nabakumar and Kapalkundala flee the ascetic in the eighth chapter of the story, titled *ashray*'e (in the refuge), in the darkness of the new moon. They slow down once they enter the forest: nothing is visible, only the starlight faintly illuminates the white sands in the distance and casts a penumbra around the tree tops (Chattopadhyay, 1953: 82). Nabakumar is led through the dark path to a Kali temple hidden amid trees. Here they find shelter, and the next morning the temple's priest marries them before leading them out of the forest and on to the path to the nearby settlement. The jungle is an intermediate space and a connecting link between the desolate coast and habitation. Darkness protects and creates the space in which to make a life-altering decision. After safely emerging from the jungle on the main road, Nabakumar discovers that what should have been a clear and safe passage is not so. Highway robbers are at work on what should have been a secure public road. Also at night, on this road he accidentally encounters a beautiful woman who turns out to be his first wife, abandoned years ago. The first conversation between the two takes place on the road in the dark. Only when they reach an inn and a light is brought into the room, does he see her: 'she is exceptionally beautiful; the beauty of her youth overflowing like the waves of a monsoon river' (*shraban'er nadi*). In a realm where open roads are not safe, the jungle provides cover; in the darkness of the night one encounters the mystery that will change the story once again.

Sudipta Kaviraj and others have noted the importance of the liminal in Bankim's novels (1995). The liminality of ethics and morals has a connection to Bankim's world-building techniques. Key encounters take place in an atmosphere of impending darkness, when the protagonist has lost their sense of direction. The sites of encounter are abandoned temples, gardens, and forts; there are frequent references to ruins overgrown with vegetation. All of these would have been packaged in a picturesque idiom by European writers and painters. Bankim refuses that aesthetic choice. Amid scenes of ruins and abandonment, moments of recognition are set apart as short durational events: lamps flicker and go out; a momentary mistake or a quick exchange of glances turns narratives. His characters are shaped in these ephemeral moments and are carved out from a landscape that is otherwise rendered in decidedly classical pictorial terms. Amid the desire to set the grandness of the historical scene, Bankim's narratives parlay in the descriptions of atmospherics and the ephemeral.

In the work of Rani Chanda, the ephemeral performs a different task. While it also helps shape characters, it is rendered distinct through a language of everyday life. Recalling a rural landscape seen through the eyes of a child, the descriptions settle on small events and everyday occurrences in a way that rarely produces a disjuncture. Liminality here is endemic and not disjunctive.

Journey toward Home

Rani's mother's village was called Gangadhar-khola; a minor settlement amid better-known villages, unable 'to raise its head'. Rani Chanda's mother's brothers, when they had moved away from the village to salaried jobs elsewhere, petitioned to have the suffix changed to 'pur'. The Bengali colloquial term for small settlement was to be replaced by the Sanskrit suffix 'pur', its classicising effect intended to make the modest ancestral village sound more respectable, civilised.

Moving away was always coming back. Returning to the ancestral home for festivities, especially during the annual puja season, was expected. Sons returning home from jobs in the city or distant provinces and daughters returning home from their in-laws' houses were much anticipated. Rani's story is a narrative of returns: on holiday visits from Dhaka, and later for a permanent stay, until they moved to Calcutta in 1927.

An accomplished dancer and painter, Rani Chanda's (1912–1997) literary career began when she started writing down the painter-writer Abanindranath Tagore's childhood stories in 1940. In Rani's writing, we find the vividness and lucidity that characterise Abanindranath's stories, as well as the pleasurable tone of reminiscence lined with the pathos of loss.[14]

Writing decades after the events, Rani needed to find access to that distant time and space of her mother's village. The village home as an ancestral place was the *permanent* home. Her mother's figure was the point of access, and the aesthetic trajectory that she had in mind needed the ordinary practices and events of village life to be articulated *in detail* in all their dynamic exuberance. The journeys that suture that narrative are the arrivals and departures of family members, the comings and goings of neighbours, and her own perambulations in that village. In terms of physical infrastructure – roads, railways, and river transportation – the village was ill-provided. Its remoteness from the administrative centres in provincial towns and capital cities was not a function of distance as the crow flies, but of the paucity of transportation facilities that move people and goods.

In the world of early twentieth-century colonial Bengal, the rural ordinary carried the double burden of colonialism and nationalism. Crudely speaking, from the colonialist perspective it was the backwaters whose only recognition was through the logic of extraction – of resources and revenue. And from the nationalist perspective, it was the domain of the folk carrying the potential of national regeneration, once it had been suitable modernised. In such imaginations, the village becomes a subject of reform and restructuring. Rani's narrative quietly disavows both these approaches. Her stories delight in the concrete materiality of place, and are alert to the possibility of alighting upon 'some little something somewhere'. (Stewart, 2007, pp. 3–4) The ordinary in sparking affect brings forth an aesthetic episode. The autographic keeps the ethnographic at bay. The memory-desire of her child-self suggests connections – more properly, digressions – without any attempt at resolution. Contrary to Rabindranath's advice to write about the good qualities of village life, Rani shows no inclination to leave out the unpleasant aspects of this village community. Her understanding of aesthetics is generous – it has no need for jettisoning the unhappy aspects of the village community. It leans toward the word's more capacious Greek derivation from aesthesis – sensation or perception – rather than its more circumscribed meaning of aesthetics as relating to beauty.[15] Finding beauty in the ordinary requires being attuned to this other possibility of aesthetics.

The kind of connections and continuities that emerge in Rani's picture of village life, however, are created not entirely through human volition. The notion of agency as individual will, presumed in the modern use of the word aesthetic, is undercut by a more complex registering of forces and sensory affects. Here 'ordinary affect', to use Kathleen Stewart's phrase, sparks connections among land, water, and humans in unexpected ways. The land and the rivers have a pace and story of their own. They disrupt and intervene. And they are accommodated.

Aqueous Aesthetic

Rani's mother visited her father's house twice a year, in summer and during the pujas in autumn, when the children were home for the holidays. The journey to

Gangadharpur from Dhaka by boat took an entire day. Rani writes in the present tense. They embark from the ghat at Buriganga and proceed toward the turbulent waters of the Dhaleswari. She gives a detailed description of how the boatmen navigate the rough waters of the Dhaleswari. The riverbanks are no longer visible: the passengers sit quietly, terrified of the muddy waters crashing against the boat amid the rhythmic yells of the boatmen. And when the boat reaches the placid waters of the Ichhamati, everyone relaxes. The boatmen cook a meal of coarse rice and fresh fish, while the family eats a cold repast. The boat is then trawled along the river's bank. In the afternoon they enter a narrow *khal* (creek/canal). The conversation between her mother and the boatmen on the river journey switches to an East Bengal dialect. Writing in English, I have no way of conveying this switch of tongues without flattening the tonality of the language and affect.

In the last stretch, the boatsmen have to get down into the shallow waters and tug the boat. When they reach the canal that joins the Ichhamati near the roots of the large banyan tree, they are close to their destination. The narrow *khal* is lined with the houses of *jola, bhnuimali, shekhs* – lower castes and Muslim families. All are on a first-name basis with her mother, and the news of her arrival reaches her home long before the boat reaches the destination. Next to the canal is Rani's grandfather's friend Baban Khan's house – Rani's mother calls him Baban-uncle. As the boat nears the Khan household, the eldest son of the family, Ali Hosen Khan, comes out to the water's edge to enquire whose boat it is. Rani's mother pokes her head out. 'It's me, Hosen-bhai'.

'Oh, it's Puni-boyindi'.[16] Her name is Purnashashi, Puni in short. Through familial addresses such as grandpa, sister, and brother, relations with neighbours are constructed as an extended family. Glad to see her, Hosen-bhai sends a stalk full of bananas and a couple of ripe jackfruits to the boat. The boat moves slowly. Rani portrays an image of an intimate community, despite differences in wealth and status: 'A daughter has returned home to the village: everyone shares that joy'.

This is a fluid landscape, transformed by the rains and *unna* or dry season. In summer, they enter the village by the *khal*, when the land is dry. By the puja season, the monsoon-fed waters of the *khal*s, tanks, and ponds become connected: then the boat goes as far as the household pond. Rani's story creates a riverine geography of interspersed spaces – *bhite, uthan, ghat* – connected by *khal*s, *beel*s (marshes), and foot tracks that disappear in the rains.

The village has two *para*s – neighborhoods of walking distance – on the east and west sides of Ghosh'er Bari'r Khal. A *bel* tree in the courtyard of the Eastern House, the house of Sushila Aunt, marks the location of the Shiva temple. Further east beyond the fields of mustard and peas is the house of Naran Bhuimali. Standing on the high bank of the pond of the Eastern House, Rani can see the roofs of the houses of this distant para sheltered by trees, but she isn't allowed to go that far. Her daily perambulations take her along the paths that lead westward. Unlike the houses on the eastern end that are separated by canals and high banks, the houses on the western end are all on one bank, and through this part of the neighbourhood stretches the path that connects the large pond of the Nandis to Ghosh'er Bari'r Khal.

With the rainy season, 'new water' arrives from the river to the *khal* that borders the villages. This is vital news, and the people react as a community:

Before dawn the news reaches that water has entered Ghosh-bari's Khal. We run to see—everyone young and old. The water advances in a torrent: the *khal* is filling up with gurgling water. The *khal* fills up. The water from the full *khal* turns south and enters the ghat of Nandi's Pukur. We run along the water. Nandi's Pukur fills up. The water takes the canalway and advances to the pond of Naya Bari. Having filled up the two largest ponds in the neighborhood, the water spills over the banks, flooding the south side and travels towards Siddheswari Tala. On the other side, the *beel* has been filling up— the waters of the *beel* and the *khal* become one.

Now the water keeps increasing: the fields, the ghats, and all the paths in the village sink under water: the banks overflow, the water approaches the boundaries of the house. And then one day it moves into the *andar mahal*— inside the courtyard of the house. . . . the water rises five fingers, ankle deep, knee deep—up to the waist and then the gurgling stops.

Grandmother hauls us onto the *dawoa* (raised porch or verandah) and makes us sit. Uncles take their books and papers up on the bed. Mother sits down to stitch a *knatha*. Aunts return to their domestic chores again. All around the calm water stands still. (Chanda, p. 39)

Much preparation has already taken place in anticipation of the rainy season. The earthen foundation of the *dawoa* has been strengthened. A temporary bamboo bridge has been suspended across the courtyard to facilitate movement among the apartments. A canoe made from a plantain trunk is used to convey goods between the rooms. Moving between households now requires boats. As the 'white water' settles just short of the raised *dawoa*, defining a new edge between land and water, the *dawoa* becomes a fishing platform and a dock of sorts.

The rains turn the rural landscape into one continuous stretch of water. Houses are built on high ground because the land remains flooded for three to four months a year. The water levels increase and recede according to an expected sequence. With the downpour, the paths between households – low lying foot tracks during the dry season – become small canals. (9)

The seasonal floods facilitate movement. The region does not have roads to speak of, and therefore travel by boat is easier than travel on foot. The school inspector visits during the rainy season, as does the matchmaker and the *shnakhari-*-conch-shell jewellery seller. The *shnakhari* moors his boat on the column of the *dawoa*. The boat's bow extends into the *dawoa* so everyone can gather around to see the boxes full of dazzling white conch jewellery.

The immersed land opens up new aesthetic vistas and opportunities. Rani accompanies her youngest uncle to the *shapla* woods that define the edge of the *beel*. They gather a boatload of these pink water lilies. The *shapla* stems are plucked one by one and peeled to prepare various dishes: *shapla* bitters, stir fry, and *shapla* fritters shaped like mini canoes. Rani weaves the *shapla* stems into garlands. (39)

As Rani's narrative moves from the dawoa to a larger dispersed landscape of households, flower and fruit trees, canals, ponds, tanks, and rivers, her readers are urged to walk with her to the sites that attracted her childhood attention. One such place is the Subachani drawn on one side of the ghat of Nandi's Pond. Painted with oil and vermillion, the three female figures, compositely Subachani (literally, delivering good words), are the keeper and conveyer of good news. This is important in a village in which there are so many close relatives who live in far-off places. Letters take time to reach. The women of the village rely on Subachani's providence to receive good news, to know that loved ones in distant places are faring well. When a postcard arrives bearing the good news of Julfi's husband's new job, Subachani is felicitated at the ghat. The sound of ululation runs through the village – a sign for everyone to gather, witness the observance, and thus share in the good fortune. (48–50) The letter is read aloud at the conclusion of the ritual in consideration of those who can't read.

Sounds of sadness and brutality carry across the village as well: every time Maroni goes to her in-laws' house she cries. Loudly. This is expected. The in-laws' house is in a distant village; it would be heartless not to cry. Her low wail travels across the fields even when the *dooly* (palanquin) carrying her can no longer be seen from the village. The wail seems to pick up the rhythm of the palanquin bearers as they carry her away. When Sasadhar Chowdhury beats his wife, her cry sounds different. So do the cries of a wife and a mother mourning their husband/son. Their cries, sudden and at odd hours, are heart-wrenching. It disturbs the composure of the village held together by the everyday materiality of sound. (49)

These details tell us something about settling on the small things of everyday life to create a joyous aesthetic as well as a deep sense of loss. Rani's recollection is predicated on never returning again. Although the stories constituted several visits, the beginning and end of the book captured the temporal and experiential limits of her acquaintance with the world of this small village. It is the temporal distance and its placement, flanked by the two large cities – Dhaka and Calcutta – that made the ordinary spaces of a rural landowning family appear luminescent with longing. If Rani's story was commemorating both Rabindranath and her mother, her narrative was built on loss in more ways than one. Unlike her mother, Rani did not have a 'father's house' to return to. Her father had passed away when she was a child. The house that he had built on the banks of the Padma as his 'retirement home', had long been swallowed by the river, and along with the house, the entire village had disappeared. Her mother's father's house filled that absence. But that sense of her mother's home as a way of remembering her mother could only be gathered together from the ephemeral moments that resist representation. To resist erasure, she latches on to the small events that comprise the fabric of the everyday. Writing in the present tense, the text sparkling with a regional dialect, she makes the events the reader's present. Removal of a temporal distance obviates the need to return – those moments endure – as if they had never ended.

Mobility and Hegemony

If Bankim's characters need to be peripatetic so their worlds may be made visible – *pratakshya* – to the reader, the everyday language in which his characters speak is contrasted with the classical alliterative *anuprash*-filled descriptions of the landscapes and the frames of a radical bibliomigrancy to bestow on his stories the longue-durée aura of history. Bankim wanted the everyday life of his present to be transformed by the magic of historical imagination.

The cadence of Rani's journey, in contrast, is of a different order. Rani's story draws its emotive force from the movement between languages – between 'proper' standard Bengali and the local language of her mother's village – that cannot be easily translated. It is impossible to convey in English the affective difference between the local tongue and the standard, urban, reformed Bengali. In the colonial context in which all Indian languages were considered 'vernacular', placed in hierarchical opposition to English, the master's language, the local languages (often referred to as dialects) were even more marginalised. Rani's use of the local tongue to convey affective encounters interrupts the hegemony of standard Bengali that had itself been formed in the crucible of colonialism and modern print culture – shaped in large measure by Bankim – and carrying a strong anticolonial strain. Unlike her uncles, who had lobbied to give the village a more respectable name, Rani figured in the colloquialisms of the everyday rural community a geographical specificity that was worth holding on to.

As I suggested earlier, perhaps behind this linguistic inheritance and insistence resides the messy, drawn-out violence of the 1947 Partition. In a region in which rivers are 'notoriously wayward', Cyril Radcliffe's hastily drawn partition boundary was premised on the assumption of geographical fixity (Chatterjee, 1999, p. 223). And as Joya Chatterjee points out, in an instance of 'astounding negligence', not only did Radcliffe not bother to understand the geography except on a paper map, neither did the Hindu and Muslim politicians who clamoured for their piece of communal territory as an imagined homeland (225). They ignored that land moved with its rivers; home was always about travel and fluidity rather than settlement and fixity.

Bankim's and Rani's narratives suggest relations between travel and bibliomigrancy in ways that help us understand the imperative to look for other ways of occupying land. If Bankim took recourse to the historical past to create what would become a nationalist aesthetic suffused with the grandness of the sublime, Rani patiently worked against this grain. In locating people and places that were all but subsumed under the category of 'folk' in Bengali nationalist imagination amid the ordinary ephemeral, she crafted a temporal space that offered potential for rethinking colonialist/nationalist projections of land as landscape. Recreating aqueous journeys helped these endeavours.

Notes

1 Historians Christopher Hill, Rohan D'Souza, Judith Whitehead, Iftekhar Iqbal, Nitin Sharma, Debjani Bhattacharyya, and others have written about the difficulties that the colonial state encountered in understanding and regulating the fluvial landscape of the Gangetic Plains.

2 The discussion of Rani Chanda's memoir is drawn from my book, Chattopadhyay (2023).
3 For Bengali travel literature see Mukhopadhyay (2002); Sen (2005); Gupta (2008), Bandopadhyay (2011) ; Dasgupta (2016), Banerjee & Basu (2015) and Harder (2020).
4 For an excellent discussion of eighteenth-century polyglossia as demonstrated in multilingual administrative and legal documents as well as poetry, see Chatterjee (2015) ; for mixed literary modes in Bengali literature, see Kaviraj. S (2004).
5 The process began in the late eighteenth century. Those works that give a sense of this vast undertaking include Roy (1995); Ghosh (2003); Bhadra (2011) and Mitra (2020). For libraries see Kabir (1979); Chattopadhyay (2022a; 2022b).
6 In novels such as *Rajsingha* he extensively uses Hindustani.
7 He was transferred 20 times during his career and served in Jessore, Negua, Khulna, Baruipur, Alipur, and Barasat in the 24 Parganas, Murshidababd, Hooghly, Habra, Jajpur and Bhadrak in Cuttack, Jhinaidaha, and Midnapore (Medinipur). *Bankim Rachanabali*, vol 1. ed. Jogeshchandra Bagal (Kolkata: Sahitya Sansad, 1994, pp. 12–13).
8 He purchased a house in Calcutta in 1887.
9 He was remembered for his role in reducing robberies in the rivers and canals of the deltaic region. C.E. Buckland wrote: 'While in charge of Khulna sub-division (now a district) he helped very largely in suppressing river dacoities and establishing peace and order in the eastern canals'. Cited in *Bankim Rachanabali*, vol 1, 15.
10 For a genealogy of the English idea of waste see Di Palma (2014).
11 The discourse on wastelands did not die off by the late nineteenth century as Vittoria Di Palma suggests; its trailing off in England was accompanied by an expansion of the idea in the colonies. The Forestry Acts in late-nineteenth-century India are critical to understand the displacement of this discourse and the contribution of the idea of waste to imperial expansion.
12 For example, the Sunderbans is referred to as 'woods infested by tygers' as well as 'thick jungles' by James Rennell in the 1760s, as 'jungle' by W.E. Morrieson ca 1814, and 'forest' with intertwined and immense trees and low brushwood. For a discussion of these see Chattopadhyay (2016).
13 For various perspectives on the specificity of the use of the term jungle to refer to tropical forests, see Frenkel (1996).
14 Rani Chanda's other works include *Purnakumbha* (1952), *Jenana Phatak* (1983), *Patheghate* (1978), *Sab Hate Apan* (1984), and the biographies, *Alapchari Rabindranath* (1942), *Gurudev* (1962), and *Shilpaguru Abanindranath* (1972). *Jenana Phatak* was based on her experience of imprisonment in 1942 for her involvement in the nationalist movement.
15 I am grateful to Arijit Sen for discussion on this point.
16 *Boyndi* refers to elder sister in the local dialect, and *bhai* refers to brother.

References

Arnold, D. (2014). *The tropics and the traveling Gaze: India, landscape, and science, 1800–1856*. University of Washington Press.

Bandopadhyay, S. (2011). A view from Bengal: Depiction of Indigenous people in a nineteenth-century travel narrative. *Proceedings of the Indian History Congress, 72*, 759–768.

Banerjee, S., & Basu, S. (2015). Secularizing the sacred, imagining the nation-space: The Himalaya in Bengali travelogues, 1856–1901. *Modern Asian Studies, 49*(3), 609–649.

Bhadra, G. (2011). *Nyara Battlalay Jay K'bar*. Chhaatim.

Browne, M. (1788). *India tracts containing a description of the Jungle Terry Districts*.

Chanda, R. (1977). *Amar Ma'r Baper Bari*. Viswabharati.

Chatterjee, J. (1999). The fashioning of a Frontier: The radcliffe line and Bengal's Border landscape, 1947–52. *Modern Asian Studies, 33*(1), 185–242.

Chatterjee, I. (2015). Women, monastic commerce, coverture in Eastern India circa 1600–1800 CE. *Modern Asian Studies, 50*(1), 175–216.

Chattopadhyay, B. (1953). Kapalkundala. orig. pub 1866. In *Bankim Rachanabali* (Vol. 1). Sahitya Sansad.

Chattopadhyay, S. (2005). *Representing Calcutta: Modernity, nationalism and the colonial uncanny*. Routledge.

Chattopadhyay, S. (2016). Traverse, territory and the ecological uncanny: James Rennell and the mapping of the gangetic plains. In K. Bishop (Ed.), *The cartographic necessity of exile* (pp. 89–109) Routledge.

Chattopadhyay, S. (2022a). A good shelf: The material culture of reading in colonial India. In C. Ashby & M. Crinson (Eds.), *Building-object: Shared and contested territories of design and architecture* (pp. 21–41). Bloomsbury.

Chattopadhyay, S. (2023). *Small spaces: Recasting the architecture of empire*. Bloomsbury Academic.

Chattopadhyay, S., Sen, A., Mukherjee, S., & Crimmel, T. (2022b, Apr 18). 'An Oddly Bookish City': Neighborhood libraries in Calcutta/Kolkata, PLATFORM.

Dasgupta, D. (Ed.). (2016). *Amadiger Bhramanbrittanta: Unabingsha Satabdir Bangamahilader Bhramankatha*. Gangchil.

Di Palma, V. (2014). *Wasteland: A history*. Yale University Press.

Dove, M. R. (2003). Forest discourses in South and Southeast Asia: A comparison with global discourses. In P. Greenough & A. Tsing (Eds.), *Nature in the Global South: Environmental projects in South and Southeast Asia* (pp. 103–123). Duke University Press.

Fazle Kabir, A. M. (1979). English libraries in eighteenth-century Bengal. *Journal of Library History, 14*(4), 436–456.

Frenkel, S. (1996). Jungle stories: North American representations of tropical Panama, *Geographical Review 86*(3), 317–333.

Ghosh, A. (2003). Coming of the book, early print cultures in colonial India. *Book History, 6*, 23–55.

Gidwani, V. K. (1992). 'Waste' and the permanent settlement in Bengal. *Economic and Political Weekly, 27*(4), PE 39–PE 46.

Gupta, J. (2008). Modernity and the global 'Hindoo': The concept of the tour in Colonial India. *Global South, 2*(1), 59–70.

Halhed, N.B. (1776). *A code of Gentoo laws, or, ordinations of the pundits: from a Persian translation, made from the original, written in the Shanscrit language*. London.

Harder, H. (2020). Female mobility and Bengali women's travelogues in the nineteenth and early twentieth centuries. *South Asia: Journal of South Asian Studies, 43*(5), 817–835.

Hoop, E., & Aurora, S. (2017). Material meanings: "Waste" as a performative category of land in Colonial India. *Journal of Historical Geography, 55*, 82–92.

Kaviraj, S. (1995). *The unhappy consciousness*. Oxford University Press.

Kaviraj, S. (2004). Two histories of literary culture in Bengal. In S. Pollock (Ed.), *Literary cultures in history: Reconstructions from South Asia* (pp. 503–566). University of California Press.

Mani, V. (2017). *Recording world literature*. Fordham University Press.

Mitra, S. (2020). *Periodicals, readers and the making of a modern literary culture: Bengal at the turn of the twentieth century*. Brill.

Mukhopadhyay, B. (2002, April). Writing home, writing travel: The poetics and politics of dwelling in Bengali modernity. *Comparative Studies in Society and History, 44*(2), 293–318.

Roy, T. (1995). Disciplining the printed text: Colonial and nationalist surveillance of Bengali literature. In P. Chatterjee (Ed.), *Texts of power: Emerging disciplines in colonial Bengal* (pp. 30–62). University of Minnesota Press.

Sen, S. (2005). *Travels to Europe: Self and other in Bengali travel narratives 1870–1910*. Orient Longman.

Shivaramakrishnan, K. (2022, March 31). *Personal*.

Stewart, K. (2007). *Ordinary affects*. Duke University Press.

Weisberg, M. F. (2015). Jungle and desert in postcolonial texts: Intertextual ecosystems. *Cambridge Journal of Postcolonial Literary Inquiry, 2*(2), 171–189. https://doi.org/10.1017/pli.2015.12

7 Moving in Circles

Chakkars in Rajasthani Women's Songs

Nilanjana Mukherjee and Gaurav Kumar

The 'circle', as a term in the conventional sense, often evokes the sense of sameness, repetition, eternity, and completeness. However, this is only true if the circle is presumed to be an abstract, isolated entity located in a Platonic space devoid of any interaction or intersection.[1] This essay intends to look at specific practices of ritual travel that can be charted as a set of circles or *chakkars* that touch each other at one or more points. Instead of the linear trajectory associated with a common understanding of travel, that is, movement, usually from a familiar place to another new one, employing a decolonial approach, we can speak of the dominant shape/leitmotif defining the form and content of many songs sung by the women of Rajasthan in the western part of India – that is, that of a circle. In a conventional scenario, not only are the women gathered up in a circle when collectively singing such songs, but the themes too reverberate the same pattern. In these songs, female desire is often activated when women are on the move, walking along cyclical routes, usually when fetching water from a source located on the outskirts of the village or grazing cattle. In this essay, we will discuss how in these songs, forms of mobilities, and thereby spatiality, are addressed in fictional and poetic expressions, focusing on the experiences and aspirations of women located in a particular socio-cultural milieu. Walking, in this case, as John Urry mentions, becomes potentially creative, contestatory, developing meaningful encounters, extending social relations, and sometimes impressing a new path (Urry, 2007, p. 72). Desire as an affect is spontaneously evoked through wandering. It carves a distinct trajectory of this movement through the surrounding spaces, routes, and moving objects. This may include pastoralists, birds, brothers, or husbands, moving along their own trajectories, seasonal or otherwise, through the desert, intersecting established circular routes of everyday, routine travel, opening up possibilities of newness and transcendence. In many of these mobility efforts, there is a sensory and embodied evocation of the home-scape, but at the same time, these women, who are usually married, express the desire to travel to an erstwhile home, for example, the maternal home, or an ideal future home that is unknown. Therefore, the immediate fractal physical mobility only triggers an imagined circuit of mobility and occasions a rethinking of the oeuvre to discover what Michel Foucault would call 'heterotopia' or managed otherness within a realm of the real (Foucault, 2002). Therefore, the ensuing discussion initiates a rethinking of the spatio-temporal logic of narratives of travel or their tellings in oral modes so far overlooked in literary studies.

DOI: 10.4324/9781003544524-11

In this sense, we are forced to complicate what we consider to be 'travel', extraneous to normative and hegemonic perceptions of travel as linear movement, which are also likely to be goal-oriented, irreversible, and largely male. In keeping with the decolonial turn addressed in this volume, this chapter will challenge a quintessentially Western hegemonic paradigm of travel by bringing a largely marginalized and overlooked practice of representing travel into the ambit of academic scrutiny. By doing this, the essay will critique and problematize Western claims of universalisation, which refuses to acknowledge differences and excludes alternate epistemic and cultural orders. In his theorization of social production of space, Henri Lefebvre, for one, has accounted for the concept of difference as a challenge to the homogenizing spatial logic of capital and bourgeois culture. However, even when he has talked of possibilities of practices governed by different conceptual determinations, he has not focused his attention on historically specific forms of difference, such as that of race, gender, or sexuality. (Lefebvre, 1991, p. 419) In the present case, the objective behind this study is to propel attention to a different paradigm of travel and spatiality, namely, from the universal to the local and even to the mental from the physical, taking gender and subjectivity into account, when we look at travel through a frame of circularity rather than linearity. Therefore, the power relations that create and sustain the hegemony of the Western framework of travel which refuse to acknowledge the realm of the non-mimetic, the esoteric, the mystic, and the sacred are hereby repudiated. We look for a fresh vocabulary to address folk and ritual practices of telling and singing, a repertoire widely found across various regions in South Asia which juxtaposes the domain of desire and play with the sphere of sentience and the more formal relations of real life.[2]

This chapter, therefore, will attempt to analyse select songs composed and sung by Rajasthani women at 'social gatherings related to childbirth, marriage, birthdays, or singing sessions organized by "'ladies sangeet'"' (Kothari, 2003, p. 156) for an understanding of spatiality, travel, and mobility employed by them. The primary texts for this purpose will be the songs collected, transcribed, and edited by the Rajasthani ethnomusicologist Komal Kothari for his interview with Rustom Bharucha in *Rajasthan: An Oral History* (2003) and by the Rajasthani author/folklorist Vijaydan Detha in the anthology *Geeton Ki Phulwari* (2014). Both these texts state that these songs have been sung by Rajasthani women 'for centuries' without dating any of them precisely,[3] while adding that even the collection of these songs has been arduous. It was executed using methods of recording and transcribing that are hopelessly inadequate to the task of capturing the complexity, literary histories, and performative contexts (e.g., collective performance, improvisation, and repetition of lines/nonsense words) of these songs (Kothari, 2003, p. 156). This is why the collection that forms the primary text for this chapter was prepared by Vijaydan Detha by focusing on the compilation of songs sung by the women of one village in Marwar, Rajasthan, and its surrounding region – Borunda (Jodhpur).

Kothari cautions us to the fact that women's songs demand a careful reading because they are often quite subtle and indirect in expression. Being non-mimetic, they are deliberately oblique in meeting the demands of social obligations, oral rhythms, and literary forms. Even then, these are 'songs from which we can learn about how they see the world, both the outer world and their own inner worlds' (157). At the same time, he points to the sociality of women's movements and worlds as he informs us that singing of songs by women has slowly come to be regarded as a 'low-caste activity' because of which men from many rural social groups have forbidden their women from singing in public while there has been a simultaneous rise in professional/public singing of 'folk' songs by women from 'middle-class and upper-caste groups' (157). Therefore, the act of singing itself is not free from social censure and surveillance in the structural domain. It is even more compelling to treat the aesthetic repertoire here as a repository of particular themes, desires, and expressions in contemporary Rajasthani women's songs, as a map of real or imagined mobilities enabled or disabled by caste and patriarchy, as fantasies of journeys to a better way of life. Many such songs remain unrecorded and evade transcription in large parts of Rajasthan, but from the performances that Kothari saw, he further observes that 'women rarely have a male audience for their songs', suggesting that they sing for themselves (158). This, in turn, hints at the pleasures involved in the act of singing as well as the contexts and activities that form the content of these songs, one of which is everyday travel by women to fetch water, graze cattle, tend to crops, or to enjoy the weather in the evening. The phenomenology and pleasures of these acts of travel, real or imagined, actual or vicarious, at times mediated through their husbands, brothers, or messengers, constitute the focus of this chapter and will be examined through sections 'Beaux and Bards', 'Birds and Bees', and 'Brothers and Bandits' that follow.

Beaux and Bards

The anthology of songs edited by Detha is divided into three major sections: *Prem ke Geet* (songs of love/desire), *Peehar va Sasural ke Geet* (songs of natal and affinal homes), and *Bhai Behen ke Geet* (songs of brother and sister). Within the available archive of consulted songs, the twin themes of desire and displacement seem to dominate. Desire necessitates movements toward and longing for new places, objects, people, and experiences. Apropos displacement, women sing about their separation from missing or migrant husbands (who may be away for jobs), brothers, or parents (from whom they were separated when they moved to their in-laws' village/town after marriage). The sense of alienation, exile, or displacement caused by the movement of the bride to her in-laws' place is so common and affectively resonant that a common word used to describe women/daughters in many of these songs is *Kurja* (as the appendix informs us), which is the local name for the migratory bird Demoiselle crane that undertakes long, arduous migrations during winter to reach the Indian subcontinent (470). In keeping with the same sense of migrancy and mobility, Rebadis or Raikas, usually glossed as shepherds, who used to deliver letters and messages before the arrival of modern means of communication, were often considered and addressed as brothers by brides waiting for any news from their parents. Their own fates and expectations were intertwined with the nomadic communities of the Raikas and their ceaseless movements.[4] Their recurring presence, movement, and significance in women's songs and their affinity with waiting women are indicative of the constant mobility and social connectedness that has historically shaped Rajasthani culture and life.

Perhaps as a result of the displacement that most Rajasthani women face after their marriage, many of the songs collected in *Geeton ki Phulwadi* demonstrate an acute awareness of spatiality. This takes many forms. When specific places are mentioned in songs about the separation of the bride from her parents, such as Jaisalmer in the song 'Mahnai kyun deeni Pardesh' (which describes a camel travelling to Jaisalmer before describing the narrator's sadness), they are used symbolically to indicate large distances of separation incurred as a result of marriage. It is important to note here that this would work only if the women singing these songs had a plausible grasp of the geography and distances involved and these places were both 'real' and 'metaphorical' at the same time. These songs also mention specific places closely related to the region these women are located in. For instance, in the song 'Loongan Hando Mehal Chunai', the court of Jodhpur is mentioned: the wife, who doesn't want her husband to leave for his job, suggests that he send his father in his stead to serve 'Jodhpur ki chaakri' or be in Jodhpur's service.[5] This song also mentions other prominent places like Chittod and Medta that her husband will pass through, places which were often traversed by Rajasthani men and had their own prosperity, glory, and charm to boast of, in an act that represents a wish for the husband's prosperity and well-being (468) as well as a circuit of desire along which women's minds often move. These places are, after all, only partly familiar to the women, which is where desire begins, as they map their personal worlds in these songs. That women in Rajasthan have possessed such understanding despite historically having limited opportunities for travel due to patriarchal constraints is borne out by another observation in the appendix that states that women's songs frequently mention famous cities and townships like Ajmer and Medta which are not *prant-vyapi* (pan-state) but *kshetra-saapeksha* (relative to the region inhabited by the singing women) (467). Kothari also tells us that not only do women's songs lack fixed beginnings, middle, and endings (subject to change during each performance), but they also often omit specific place/ character names and work with general situations that allow the singing women to 'use her imagination and enlarge the narrative space' (164) by inserting the places and names close to their location and culture. As a result, if songs from different regional communities and social groups of Rajasthan are compiled and juxtaposed, they should reveal cultures of imagined maps of subjective regions that inform the lives of Rajasthani women.

The other form in which women's sense of spatiality is expressed in these songs is in their awareness of the networks of trade and mobility that have historically crisscrossed the entirety of Rajasthan, stretching to regions like Central Asia in the North and several parts of South India in the South. For instance, *Heena* or *mehndi* from the Malwa region as well as Amarkot (now in Pakistan) finds mention as a desirable object in songs like 'Mehndi nipjai Malwai'[6]: '*Mehndi upaje Malwe aayi Umarkot*' (*mehndi* from the Malwa region has now reached Umarkot for trade) (247). Moreover, this *mehndi* is prepared for application by mixing it with the water from the river Ganges, which was often brought to Rajasthan by traders as well as local pilgrims returning from a pilgrimage outside the state. In the song 'Meethi Vaani bol', the salt from Sambhar Salt Lake, which was often supplied to various parts of Rajasthan by nomadic pastoralists, is declared bitter in contrast with a mother's speech by a woman seemingly engaged in a conversation with a *kauva* or crow (345). References to the cattle trade prevalent across the Thar desert are made in songs such as 'Gujaratan todadli' and 'Sovan moriyo' by mentioning 'Gujarati' or *Kuchhi oontni* (camels from the neighbouring state of Gujarat) owned by family members (usually brothers, as evident in 'Sovan moriyo' when the narrator asks which of her brothers will ride a horse and which one a camel from Kuchh) (339, 393). At the same time, there are figurations in such songs such as the mention of the *saat salaam* commonly used across Rajasthan,[7] presence of generic character names like Narayan Das,[8] and references to flora and fauna typically associated with Rajasthan, such as the *Ker,* as well as familiar objects like trinkets used to decorate camels familiar to most Rajasthani people. This implies that women's songs are imbued with a geo-consciousness that includes both the sense of Rajasthan as a region imagined as united by a common language, culture, and history as well as a region diversified along various vectors (land, vegetation, language, etc.). This unity and regional bonding arose from an idea of a distinct ethnic conception of community based on reasons of war and a sense of the place from which the singing women sing of their world. These songs also juxtapose the real environment of inhabitation with fantastic and fabled lands outside the familiar, giving rise to a unique imagined cartography of a homeland.

As mentioned earlier, many of the places and routes of travel alluded to in these songs signify desire, which in its essence is an orientation or movement toward the other, which could be a place, a person, or a life. Often, places are named casually and almost indiscriminately, signifying a whole constellation of desire of the life, the luxury, the spaces, and people they may represent. Kothari also remarks that for songs without a clear narrative, the stanzas are more or less independent, leading to 'jumps' between the stanzas, mimicking the movements of an aimless wanderer interested not in a particular destination but the pleasures of movement alone. This is reflected in the swift change in scenery from one line to another, one stanza to the next (163). In many of these songs, the movement is so seamless that there are no stanzaic divisions to be detected during the oral performance or the transcription process but are inserted later for the ease of the reader. Is this a desire for specific places? We can surmise without hesitation that this signifies a restless aspiration, an ecstatic energy that constantly keeps one moving and seeking, from one object of desire to another, from one place to another. In certain songs, this desire takes specific forms. For instance, the songs 'Moriyo' (57) and 'Dekho in Jogi rau Roop' (43) tell the story of a married woman who visits the village pond with her sister-in-law to fetch water and is immediately attracted to a peacock (in the former) or *jogi,* a wandering mendicant (in the latter). The following is our translation of the first few lines from 'Moriyo':

O Moon, your moonlit night,
Two sisters-in-law set forth to get water

Off they went to the big pond
Placed the pitcher on the bank
And the pitcher rings upon the frangipani bough
Wandered and surveyed the gardens
Plucked tender stems for brushes
Scrubbed their feet clean
Scrubbed the ivory bracelets shiny
Scrubbed their hands clean
And cleaned their teeth until they shone
A peacock sits at the edge of the pond
Spreading its wings to cover the pond
'O Indra's Peacock, step aside a little
Let us fill our pitchers with clear water'
'Take off the veil, O beauteous maid
Before you fill your pitchers with water'
'This veil is not to be lifted, O Peacock
This veil lifts only in the pleasure palace
This veil lifts only in the cloud palace'
'Behold, sister-in-law, this peacock's beauty
Twice as stunning as your brother's' (57; translation ours)

In 'Dekho in Jogi rau Roop', the married woman exclaims: 'Behold, sister-in-law, this jogi (mendicant)'s beauty/Twice as stunning as your brother's!' (43; translation ours). In both songs, the third line ends with *bade talab*, telling us that they went to the big pond. The lineation naturally ends up highlighting the word *talab*, usually employed to refer to ponds or lakes in Rajasthan, but here, the word obviously suggests the distance traveled away from society to reach the pond. These ponds were usually located at strategic locations near the boundary of the village in order to collect the rainwater runoff from the surrounding hilly areas or highlands. (Thilak, 2019). The liminality of such sites located at the boundary of a village, settlement, or civilization offered women a gathering place, where they could meet and converse freely with people, irrespective of their gender, away from the restrictions of their homes. Moreover, the theme of wandering, such as in the line, 'wandered and surveyed the gardens', further suggests that when women visit these water sources, especially on pleasant, temperate, moonlit nights, they also engage in leisurely loitering, pursuits, and conversations that are not necessarily restricted to their purpose of visit, that is, fetching water.

These two songs are identical in the story, and most lines are repeated verbatim, except for the substitution of the mendicant with a peacock, and a more gruesome ending in 'Moriyo'. In the latter, the woman's husband sadistically serves his wife the peacock's meat for dinner, instead of simply killing the peacock/lover, which is what he does to the *jogi* in 'Dekho in Jogi rau Roop' (46). Curiously, in both versions, the object of desire is a wandering entity, or perhaps it is the desire for wandering itself. At one point in the song(s), the sister-in-law informs her brother that his wife has traversed 'the mountain' to be with her lover. The note attached to the line tells us that the mountain referred to here is the Aravalli, which separates the more arid regions of the Thar desert from the eastern parts of Rajasthan consisting of plains and plateaus, and the verb used here, *laangh*, has the connotations of a dangerous cross over that involves transgressing societal norms (60). Interestingly, the woman would usually be aware that her sister-in-law would report any transgression against marital fidelity and yet confesses her fascination for her object of desire, perhaps carried away by an impulsive desire that seeks some expression, some action, and mobility. The same desire, turned into anger, later makes her do things that will be sacrilegious for a married woman. She violently throws the bowl she was holding and immediately orders her maid to get the peacock painted on her veil, its shape carved on her head jewelry, and its image tattooed on her eyes. The inscription of the peacock motif on the veil, the *rakhdi* (gold head-ornament, worn by married women as an auspicious mark of their marriage), and the eyes, represent outraging the auspiciousness, fidelity, and modesty of marriage. In both these songs, transgressive female desire is activated when women are on the move, to fetch water. Wandering itself figures as an affect or object, reflecting on and carving a trajectory of desire through the spaces and regions that surround them. The Aravalli mountains mark a watershed between one's immediate space of limited access to circulation and the otherwise limitless world of transgressive desire. Unlike, say, a beggar or bandit who has been outlawed or banished, the *jogi* is a 'legitimate' wandering figure, whose mobility is seen as permissible or even venerated within the framework of modern state and society. His figure is, therefore, the most appropriate for activating and catalyzing a desire for aimless motion as well as symbolic of the vagabond-outsider's freedom, offering a 'fuzzy' perspective on/ critique of a sedentary and normative regime of socio-moral decorum (Cresswell, 2011, p. 247).

The song 'Raydhan', wherein the narrator of the song meets Raydhan, the eponymous handsome man, when she goes out to fetch water and subsequently describes his sensuous appearance to her mother in the hopes of convincing her to get them married, is another reference to the possibilities of desire that such everyday ritual travel inaugurates:[9]

> …I went to the pond to fetch water
> Where Raydhan came riding on a dashing horse…
> His neck adorned with a *Bhairav* rosary
> A *Maimadshahi* turban on his head…
>
> I went to fetch water
> Where Raydhan came riding on a handsome horse (88; translation ours)

The transferred epithets used for the horse, 'dashing' and 'handsome', qualify their real object, the handsome Raydhan, whose attractiveness is enhanced by stories about him that circulate and make him popular among women like the narrator of this song. His accessories like the '*Bhairav* rosary, *Maimadshahi* turban', as well as the figure of the horse, which has been historically associated with royalty, might, and power as few elites could afford them earlier, compared to camels which were locally and cheaply available.

> Another song in the collection 'Piniyari', placed immediately after Raydhan, confirms the regularity of such encounters by disclosing that the man the narrator met on the way to the pond was her husband, out to test her fidelity.
> I met a camel rider, mother-in-law
> He asked me about matters of the heart
> Who was he lanky like *panihari* (woman bringing water)
> Whose face did his face resemble, dear
> He was lanky like my brother-in-law, mother-in-law
> He had a face like that of my sister-in-law, love
> This one must have been your husband, *panihari*
> He must have been out to test you (92: translation ours)

The fact that the woman in 'Piniyari' approached this man first and asked him for help with the water pots suggests how dangerously close to transgression women could come while they were out to fetch water, even without it being considered anything out of the ordinary.

In another very interesting *Ghoomar* song, 'Othido', included in a collection published by Rajasthan Sahitya Akademi (1972), the husband of the woman singing the song is a herder of camels who has been away from his village for many years. One day, the woman notices a flock of camels passing through the village. She mistakenly assumes that one of the men herding the camels is her husband and joins their group. When the man starts moving suspiciously fast, the woman realizes that she is not with her husband but another man. In most versions of the song, she kills herself when the man doesn't stop. What is interesting for our discussion here, however, is the fact that the woman decides to join a group of men and camels at all. If it were her husband returning home, why would he not stop after reaching his village or home? More importantly, did the woman really want them to stop, or did she simply want to join them?

> A procession of camel riders is passing through
> My *Maru's* procession is passing through
> Which of these camel riders is my (elder) brother-in-law
> And which one of these riders is my husband
> The one seated in the front is my brother-in-law
> The one sitting behind him is my husband
> The one wearing a *moliya* (a turban style) is my brother-in-law, and the one with a *Pecha* (another type of turban) is my beloved
> The one with *murakiyan* (small earrings worn by men) in his ears is my brother-in-law, and the one with *murakiyan* and *saankalen* (another specific type of Rajasthani jewelry) is my love
> The one with bhanwariyan in his ears is my brother-in-law, and the one with many rings is my dearly beloved

> The one with the bracelets is my brother-in-law,
> and the one with a *kanthi* (a necklace with basil seeds on a string, usually
> worn by some Hindu devotees) is my darling (100; translation ours)

The sensuous and detailed description of the camel riders, coupled with the romantic associations invoked by the allusion to Rajasthani love ballad of Dhola-Maru (in which Dhola, the prince of Marwar, famously leaves his wife and kingdom behind to secretly reach Maru's place in Marwar at a crazy speed that covers the great distance between the two kingdoms in a day and half a night) remind us that temptations of desire can overcome the obstacles of speed and inhospitable terrains. The examples discussed above begin to sketch the circle of desire opened up by women's movements to and through the village pond, meadows, roads, and fields, where they encounter other movements that arrive and intersect from the far-off unknown.

Birds and Bees

Many of the songs recorded in *Geeton Ki Phulwari* contain references to birds and animals, often addressing or describing them with playful words. The rhythms of walking that Rajasthani women experience during their everyday travel open up avenues of living, experiencing, and feeling, allowing them to immerse themselves in a sensorial realm of tempting movements manifested by animals and other elements of their environment[10]. In the essay 'Walking: New Forms and Spaces for Studies of Pedestrianism', Hayden Lorimer reminds us that 'walking does not come naturally to all and is not embodied in exactly the same way' (24). This essay points out that sauntering, marching, sleepwalking, tiptoeing, or walking as a blind person are all very different experiences. In the aesthetic space, they reflect a variety of emotional states. Walking as a Rajasthani woman with a waterpot, harvested crop, or leash in hand while wearing traditional clothing (e.g., the loose *saree* and skirt that are worn by women in many parts of India) has its own peculiar haptic, visual, and spatial rhythms that differentiate her walk as compared to those of her fellow travellers she meets on the road. Her own constraints contrast with the apparent freedom of the birds, men/*jogis*, etc. who are dressed and move more uninhibitedly, just as the wind and the grass that it propels with impetus and pleasurable gusts. Lorimer proposes the idea of a 'desire line' (for instance, the trail marked by the footfalls of previous walkers on a hill) that is dictated both by the needs and moods of those who have walked before as well as the shape and resistance of the landscape (28). This may be why we often see Rajasthani women singing songs, imagining the experience of wandering birds (*tota, maina, kuraj,* etc.), animals, and *jogis* that travel parts of the globe that are destined to remain inaccessible to most women in those contexts, including the natal home or the city where the husband works and lives alone. Walking and flying are linked to an idea of transcendence from the immediate reality (of repression) to a better (utopian) one and therefore is likened to one of opportunity with special reference to gendered immobility pervasive in the culture. If we gloss over the affective textures in such fictional renderings as unreliable, we will do injustice to their underlying aspirations of rights, equity, and equality of a specific gender present in such collective oral voicing.

In the song 'Meethi Vaani Bol', for instance, the narrator seems to be engrossed in a conversation with a crow who is addressed as the interlocutor for a series of questions posed by the narrator. These questions concern the empirical qualities of events and objects in the world that may be of interest to the narrator – 'What do the river waters look like?', 'What do *Rohida* flowers[11] look like?', 'What does the foam of the ocean look like?', and 'What do the cotton flowers look like?' (344). However, what is interesting is that the answers are also supplied by the woman herself (as evidenced by the term of address *kauve* or crow that punctuates each of the questions as well as answers). Thus, the woman seems to be able to describe river waters, ocean foam, and the clothing, gait, and the camel owned by her brother as he leaves her natal home to meet her at her in-laws' house. This seems physically impossible since most Rajasthani women would not even have travelled enough to describe ocean and river waters except during a pilgrimage to a coastal town, a rare journey undertaken by a few affluent women. The only explanation that makes sense here (without making the presence of the crow redundant by reducing it to a mere refrain that punctuates scenes imagined by the woman) is that the woman actually imagines the flight and aerial view experienced by the crow as it flies through the world that has become the object of curiosity and desire for the narrator. The focus on colors and visuality, mixed with the woman's own experiences (the slightly bitter taste of *Sambhar* salt, the strong taste of unripe *Ker berries*, etc.) seem to confirm this intermingling of perspective and senses thereby giving rise to an interesting phenomenology of movements.

In another song titled 'Kukrau' or 'Rooster', which has been interestingly placed within the section *Prem ke Geet* by Detha, the singing woman presents passionate entreaties dedicated to the bird, inviting him to visit her courtyard. This opens up the possibility of an alternative reading in which the woman could be assumed to be making these arguments for her paramour. The magic realist texture of the song and its placement in this section by Detha together make it inescapable to be read as the woman's overtures to a reluctant roving lover.

She addresses him as *bharpur jawan* or 'ripe youth' and offers cardamom as feed.

> …You are a blossoming youth now
> O red rooster, I shall scatter some cardamom for you in my courtyard
> Come to my courtyard for some pecking (132; translation ours)

When the rooster complains about her courtyard being too rough on his beak or when the woman's mother-in-law or sister/brother-in-law abuses the bird, the narrator first distracts him by promising to buy him anklets, which he can use to come dancing and strutting, and then reassures him by reminding him that the mother-in-law is old and will die soon, the brother-in-law has to go away for his job, and the sister-in-law is herself like a bird, destined to leave soon once she gets married. The song presents another instance of a woman beginning to experience the ground or the sounds in her surroundings as experienced by a critter (the feel of the ground against the beak is a peculiar detail) and entertaining herself with the thoughts of his trajectories of movement, hoping that it crosses her house, bringing her some joy, novelty, and extraordinary company outside the regimented ambit of her affinal home.

Brothers and Bandits

Songs such as 'Kuraj Juraive Saasrai' (414) remind us of the intense desire for movement between the natal and affinal homes that characterizes the subjectivity of a married woman. Since most women are only allowed to visit their natal homes on specific occasions, this desire to travel or return is transferred onto the figures of their brothers in many of these songs. We notice that brothers often become the bearers of a misplaced aspiration because of their ability to travel without restriction and to bring tidings of their travels as well as news of the folks back in her maternal home. This song, in particular, begins with a woman calling herself a *kooraj* (for the reason mentioned earlier) and requesting another bird, such as a parrot or a swan, to go to her brother with her message because of her inability to travel, inviting him to her in-laws' place. The brother complies to her utmost delight. Whenever the brother visits, fascinating conversations and tales follow. The song 'Moolmool Aachhi hai' suggests how pleasurable and fantastic these tales of the brother's exploits and travels can be. The song begins with a curious woman excitedly mentioning details of the robbery that her brothers may have committed in Gujarat, confirming and celebrating the feats of her younger and elder brother (400). Tanuja Kothiyal's book *Nomadic Narratives: A History of Mobility and Identity in the Great Indian Desert* (2016), which delves into the history of mobility that defines Rajasthan, is a reminder of 'the mobility of the Thar's inhabitants, who were warriors, pastoralists, traders, ascetics and bards, often in overlapping capacities, exchanging mobile wealth and equally mobile narratives' (1). This book, which 'aims to recover a history of mobility and multiplicity of narratives, authorities, and boundaries against the thrust of conventional history of Rajasthan, wherein land, agrarianism, territorialism, and rigid identities and boundaries defined by/around Rajputs dominate', makes it clear that raiding and banditry were a way of life for many groups and castes living in the Thar region before it was slowly removed from the domain of acceptable practices for most communities (Kumar, 2021). The song goes on to excitedly describe the *Kutchchi* camel and horse that the brothers will ride for various journeys (bringing them *chunri, odhna*, etc.):

> Which of my brothers will ride the horse, mother
> Which one will ride the *Kachhi* camel?
> The elder brother will ride the horse, mother
> The younger brother will ride the *Kachhi* camel

> This *Malmal* is nice
> How beautiful are the colors of this *odhni*, mother!
> Which brother got me this *chunari*, mother
> Which brother the green *odhna*? (translation ours)

Another interesting song 'Bhawaj Hejla Saiyaan ai', (357) describes the rare occasion when a woman is allowed to travel to her brother's place. On the way she meets a camel rider, a milkman, an astrologer, and a maid with whom she initiates conversations to confirm the way to her natal home.

> On my way, I met a camel rider
> Tell me the way, O camel rider, to my natal house, love
> The trail on the left goes to Dwarka, my sister
> The trail on the left is that of your house, sister
> On the way, I met a milkman
> Tell me the way, O milkman, to my natal house, love
> The trail on the left goes to the farmlands, my sister
> The trail on the left is that of your house, sister
> On the way, I met Rambha, the maid
> Tell me the way to my father's house, Rambha
> Grand foyer, many windows
> Open space in front of the house of your father, sister
> On the way, I met the astrologer Joshiji
> Please tell me the way to my father's house, Joshiji
> The open space in front of your father's house lies straight on the sun's way
> The bananas dancing in the wind are your father's, sister (translation ours)

Incidentally, she is told that the brother is away on duty and is, therefore, unable to meet him before her return (360). Therefore, the sense of home is not embodied in one single person but diffused. In the appendix to the book, Kothari makes an observation that it couldn't possibly be the case that the woman actually forgot her way home. She simply wanted to hear people talking about her folks and home (446). One may add to this account her potent interest in meeting and chatting with fellow travellers so that she could learn about their lives, movements, and journeys. One cannot, after all, help another person navigate without having a prior fair knowledge of the terrain. The song thus explores a relational dimension that cannot be simply reduced to the materiality of the place called home. Playing on the idea of self-achievement and self-recognition through conversations while on the move, it explores in the same mobile way not only the relationship between self and space but also that between community and space. It, therefore, provides an instance of connected lives and of co-creation of space through relationships with neighbours and neighbourhood. Therefore, this connection makes the process of place-making organic, cyclical, and circular, not limited to the locus of the self alone.

Conclusion

As can be inferred, trajectories charted out in these songs are nourished by undefinable sources of energy and are inspired by amorphous complex of contents and potentials which arrive from all directions. The utterances and journeys of the Rajasthani women exist for and with one another and through an exchange of stimulus. Their singing encourages us to think about their space not in the modern sense of a cartographic logic of fixities but as fluid, connected, and shared. The centrality of vision that largely dominates modern geo-consciousness is anathema to the composite nature of the sensorial order which manifests a world of longing and belonging. In this sense, these renditions indicate the making of a life-world, which is not marked by egotist implications of power. We could term it 'experiential space' as Henri Lefebvre would have, but would rather speak of these practices as creating a deeply felt space instead: one which emanates from bonding and exchanging with all and every kind of life form available in the immediate surroundings.[12] This geo-consciousness is nebulous and far from being pragmatic, utilitarian, corrective, or knowledge propelled. We, therefore, return to the centrality of the circular or the spiral, as opposed to the linear, as the defining form that constitutes the experience of Rajasthani women's travels as reflected in their songs. Kothari tells us that the inherently collective nature of women's singing allows for situations wherein married women, whose husbands live with them, join in songs of separation while unmarried women sing songs about in-laws. With this association, everyone is turned into equals providing a saving emancipation from the gloomy persistence of purpose-driven realities of life. Therefore, even if these songs tend to reflect the interiority of individual singers, they are not so, for they forge a realm beyond the individual, which is shorn of all personal features. This enlargement of experience/emotion/subjectivity by including others' experience, of course, reminds us of the metaphor of the chakra or wheel (or a flowing river that begins and ends in a circle) that is used to explain the Mahayana Buddhist doctrine of inter-dependent origination, in which everything in the universe is connected to and dependent upon everything else like a 'house of cards', ensuring that there are no unchanging or isolated substances but a 'stream of causally connected qualities' (King, 1999: 82)[13]. In this sense, the journey or *chakkar* taken by the local woman in Rajasthani folk songs from her house to the pond is endless because it opens up ever so slightly, the paths and pleasures experienced by the bards, brothers, and beaux she meets on the way.

Notes

1 In several of his dialogues, including, most famously, the *Republic*, Plato outlines a metaphysics which posits a dualism at the heart of reality – the world of worldly, material things, and the world of *forms* (e.g., the form of beauty and circularity), the source of all worldly things. Forms, assumed to be eternal and perfect, are not supposed to exhibit properties that might characterize worldly things such as change, relationality, and contingency. This space of Forms, therefore, is an ideal one, wherein interactions

and changes for entities/Forms, for example, circles, that are common occurrences in the material world are forbidden.

2 The term 'play' is specifically helpful here if we approach the songs from the reference point of 'sociability' as talked about by Georg Simmel in his essay 'The Sociology of Sociability.' According to him, human associations or sociability can occur when serious purposes of individuals are kept out and whose deeper nature can appear more integrated if we do not try to impose and comprehend it realistically.

3 In the appendix to *Geeton Ki Phulwari*, Kothari mentions that songs sung by tribal men and women in Rajasthan on the other hand contain contemporary allusions to the arrival of railway, wars, new laws, etc. and thus can probably be dated more accurately than women's songs that contain stories and scenes that often use generic names/descriptors for people, emotions, and places (439).

4 The Raikas are historically a very important part of Rajasthani life and culture wherein these nomadic peasants and pastoralists formed a 'precariously balanced system' to which the pastoralists would bring essential messages/letters, trading products such as butter, tools, ghee, cattle, salt, and hides. (apart from the mandated taxes) and received stubble or grass for their cattle, and food, water, as well as a place to rest in themselves. (Kothiyal, 137).

5 The appendix tells us that this is because Jodhpur was one of the prominent Rajput princely states (along with Bikaner, Udaipur, and Jaipur, which are also frequently mentioned) or *rajwadas* where the husbands of the women singing these songs often worked and lived, away from their home and family (113, 468).

6 Tonk-Toda, which is close to Udaipur, is mentioned as another place famous for its mehndi in the song 'Kiso Viro aavai malekto' (425).

7 Literally translated as seven salutes, a greeting that the appendix tells us is specific to Rajasthani songs.

8 The appendix tells us that a variant of this name is popular across regions such as Jodhpur, Bikaner, and Kishangadh.

9 In Detha's 'Him-Samadhi', his version of the folktale popular in Western India, Sheni, the daughter of a Charan, the headman of a village, similarly meets and falls in love with Vijanand, a roving singer and cowherd who doesn't even own the cows he herds for a living, thus threatening the stability of a sedentary, agrarian way of living with suggestions and temptations of itinerancy and indeterminacy (Kumar, 2021). As cross-reference one can cite numerous songs on Radha's trysts with Krishna on similar occasions of fetching water from the river Yamuna.

10 Hayden Lorimer (2011) points out that:
    rupture caused by the industrial revolution – 'when walking ceased to be part of the continuum of experience and instead became something consciously chosen.' (Solnit, 2000, p. 265) – was not experienced universally or evenly. For indigenous groups, walking is central to existence, and a continuous means to learn and to teach place-knowledge.

11 It is a flowering tree (*Tecomella undulata*) commonly found in the Marwar region of Rajasthan.

12 Lefebre talks of a triad of spatial practices through which a given space can be socially produced. However, most Western theoreticians like Lefebvre have emphasized on bodily practices which mark and create space.

13 In Advaita Vedantic thought, something similar is suggested by asserting that nothing is isolated, and everything is connected to everything else because it is all one – *Brahman*.

References

Cresswell, T. (2011). The Vagrant/Vagabond: The curious career of a Mobile subject. In T. Cresswell & P. Merriman (Eds.), *Geographies of mobilities: Practices, spaces, subjects* (pp. 239–254). Ashgate Publishing Limited.

Detha, V. (2014). *Geeton ki Phulwari* (K. Kabir, Trans.). Sahitya Akademi.

Foucault, M. (2002). *The order of things: An archaeology of the human sciences*. Routledge.

King, R. (1999). *Indian philosophy: An introduction to Hindu and Buddhist thought*. Edinburgh University Press.

Kothari, K., & Bharucha, R. (2003). *Rajasthan: An oral history*. Penguin Random House.

Kothiyal, T. (2016). *Nomadic narratives: A history of mobility and identity in the Great Indian Desert*. Cambridge University Press.

Kumar, G. (2021). *Alekhaan Prem: Rethinking love and translation through Vijaydan Detha's short stories* (dissertation).

Lefebvre, H. (1991). *The production of space*. Blackwell.

Lorimer, H. (2011). Walking: New forms and spaces for studies of pedestrianism. In T. Cresswell & P. Merriman (Eds.), *Geographies of mobilities: Practices, spaces, subjects* (pp. 19–34). Ashgate Publishing Limited.

Purohit, M., & Vyas, M. (1972). *Ghoomaren*. Rajasthan Sahitya Akademi.

Solnit, R. (2002). Wanderlust: A History of Walking. New York: Viking.

Thilak, N. (2019). Water architecture of Rajasthan: A journey through Jodhpur and Bundi. In *Sahapedia*.

Urry, J. (2007). *Mobilities*. Polity Press.

The Trope of Homelessness

8 Homeless in Gujarat and India

On the Curious Love of Indulal Yagnik

Ajay Skaria

In the 1920s, the writer Ramanlal Desai (the father of the Marxist sociologist A.R. Desai) drew a comparison between two Gujarati figures whom he considered among the most prominent in the post-Gandhian generation – K.M. Munshi and Indulal Yagnik. He described them as symptomatic of two strands of Gujarat: Munshi was a *pratibashaali siddh purush,* a resolute and accomplished man, and Yagnik an *asthir man na fakir,* a mendicant of unstable mind (Yagnik, 1955, p. 8).

The remark is reproduced by Yagnik – without too much comment – in the preface to the first volume of his *Atmakatha* or autobiography. Indeed, the six-volume autobiography is structured almost as an analogy to the remark. It draws heavily on the trope of homelessness, a state often associated with *fakirs.* But this is a distinctive kind of homelessness – it is accompanied by a love of having a home. There was the homelessness produced by his vow to remain celibate, first made in 1914–1915 when he was 23, and broken briefly a few years later because of family pressure to accept the marriage performed when he was a child. In 1923-1924, just before he decided to separate from his wife and renew his vow, 'the deep, inchoate thirst for love continued to harass me...'. 'A yearning for a life filled with love in its many colours started burning in my heart' (Yagnik, 1956, p. 263). But finally, after the advice of an unnamed friend, he decided that he 'should leave the *maya* of a home (*ghar*) and remain a sacrificing person without a home – the attraction of even the dearest person should be reduced to ashes, that is all' (Yagnik, 1956, p. 265).

There was also the homelessness that accompanied his love for the *khedut* or farmer, and *mazdoor* or worker. Indeed, he differentiated himself from Munshi primarily in terms of this love: 'He was a great admirer of leaders like Caesar and Napoleon, of Dayanand Saraswati and Aurobindo Ghosh, whereas I was wedded to the common people' (Yagnik, 1955, p. 179). Visiting Panchmahals in Gujarat in 1918, he reported that he saw:

[Y]esterday's miserable looking Bhils had worn good clothes, and were moving about joyously.... As evening fell, the Bhil people's merrymaking increased. When, in all the large open grounds around town, I saw hundreds of young boys and girls holding each other's hand, singing and dancing in circles to the rhythms of the drum, my heart swelled with love and pleasure.

DOI: 10.4324/9781003544524-13

> An enthusiasm rose in my heart to serve such innocent children of mankind
> all my life.
>
> (Yagnik, 1956)[1]

And yet this love for common people also led to homelessness in relation to both
the common people and to mainstream political parties and projects. Yagnik began
with Gokhale's Servants of India Society, went on to a more radical wing of the
Congress, then to Gandhi, and then left politics for journalism. He then turned to
organizing peasants and, with the founding of the Kisan Sabha, arguably emerged
as among the three most prominent leaders of peasant struggles in early and mid-
twentieth century India. He came back to the Congress briefly after independence
before leaving it again and starting the movement against it for a separate state of
Gujarat. During this period and later, he also ran an *ashram* on the banks of the
Vatrak river and became involved with trade unions in Ahmedabad. Reflecting on
his life in one particularly dispirited moment in 1956, he wrote:

> I feel upset with myself. I have chosen to lead only unsuccessful fights. Now,
> I have reached the final point. I am regarded as unintelligent and unfaithful.
> When I cast my glance on 64 years the result is a big nothing. I go to meet
> several friends but rarely do they come here [to his *ashram,* in Nenpur, near
> Ahmedabad] to meet me. It is as though nobody thinks of me at all. Because
> of this, I sometimes feel tired of life. It affects one's self-respect to go aim-
> lessly to the city [Ahmedabad], knocking over and again on people's doors.
> When nobody has any need for me, why should I keep moving back and
> forth.
>
> (Yagnik, 1973, pp. 18–19)

Even if not usually stated in such dispirited terms, this was again a theme in his
Atmakatha: that he had never been quite at home in any of the major political
parties.

If we begin – and, as historians, we surely should – with the assumption that
homelessness does not mean the same thing in all times and places, and that metaphors, too, have history, then the question follows: how was Yagnik's homelessness produced? Within what conceptual field, more precisely, were Yagnik's
simultaneous love of home and homelessness conceived? Here, I would like to
situate Yagnik's simultaneous love of home and homelessness in relation to two
other sets of practices of home and homelessness – those involved in the logic of
transcendence and the politics of neighbourliness.

The logic of transcendence, I would like to argue, was characteristic of mainstream nationalism. In one of the most often quoted passages in Nehru's *Discovery
of India,* he remarks:

> Sometimes as I reached a gathering, a great roar of welcome would greet me:
> Bharat Mata ki Jai – 'Victory to Mother India'. I would ask them unexpect-
> edly who was this Bharat Mata, Mother India, whose victory they wanted?

My question would amuse them and surprise them, and then, not knowing exactly what to answer, they would look at each other and at me. I persisted in my questioning. At last a vigorous Jat, wedded to the soil from immemorial generations, would say that it was the dharti, the good earth of India, that they meant. What earth? Their particular village patch, or all the patches in the district or province, or in the whole of India? And so question and answer went on, till they would ask me impatiently to tell them all about it. I would endeavour to do so and explain that India was all that. they had thought, but it was much more. The mountains and the rivers, and the forests and the broad fields, which gave us food, were all dear to us, but what counted ultimately were the people of India, people like them and me, who were spread out all over this vast land. Bharat Mata, Mother India, was essentially these millions of people, and victory to her meant victory to these people. You are parts of this Bharat mata, I told them, you are in a manner yourselves Bharat Mata, and as this idea slowly soaked into their brains, their eyes would light up as if they had made a great discovery.

(Nehru, 1981, p. 60)

Notice the way that India as a nation-home is configured in this argument. On the one hand, there is the vigorous Jat wedded to the soil for immemorial generations. This Jat already is India, he already is the nation-home – it is this fact which makes India possible. On the other hand, there is also a gap between this Jat and Bharat Mata – the two have to be constantly brought together in a fusion; the Jat has to realize himself by transcending his 'Jat-ness,' so to speak, by subsuming it within Bharat Mata. Much of the work that nationalist thought set itself involved bringing about this transcendence; hence, of course, the repeated emphasis on building patriotism and national awareness.

This logic of transcendence does not reject or disregard the very local and particular homes represented by the Jat, the peasant, and other such figures. Rather, it affirms them in a particular way, that is, by focusing on how they are transcended into higher levels of generality and abstraction; the insistence is that it is only through such transcendence that the true meanings of the local home are realized. It is in this fairly precise sense of transcendence that mainstream nationalist thought can be described as cosmopolitan. This is not so in the sense that nationalist thought seeks to embrace or subsume humanity within it (quite the contrary, obviously); it is so, rather, in the sense that it is constituted by a logic that subsumes identities designated as local, contingent, or particular within more general identities. This process created both the nation-home and a certain homelessness in relation to those particular identities – the Jat and the peasant – which had been transcended. Much scholarship on nationalism has been characterized by the same logic of transcendence. Thus, for instance, one crucial innovation of Anderson's *Imagined Communities* lay in specifying, very precisely, the way in which such transcendence occurred. For Anderson, this transcendence occurs through an emphasis on the 'empty homogenous time of history that allows all the different parts of nation to exist all at once in some nationalist imaginary of simultaneity' (Anderson,

1991). In this analysis, the characteristic technologies of nationalism – the map, the museum, the census, and, of course, print capitalism – represent efforts to translate the nation-home into the terms of abstract time and space.

Another kind of home and homelessness was produced by the politics of neighbourliness. Gandhi was one of the most articulate practitioners of this politics in the twentieth century. This politics, too, claimed India as a home, but it was in a tense – and even perhaps conceptually incommensurable – relationship with the nation-home constituted by a logic of transcendence. It claimed the immediate neighbour or *padoshi* as home, and insisted that the nation was constituted not by transcending the singularity of the immediate neighbour, but by serving that singularity. I do not wish to suggest that this politics affirmed some authentic local identity, some substantive modern ethnographic space autonomous of, or prior to, mainstream nationalism. Such modern spaces or identities do not exist, and any attempt to describe or recuperate them would only reinscribe those practices that constitute mainstream nationalism. Rather, I wish to argue that the politics of neighbourliness – and especially Gandhi's version of it, influential in early twentieth-century India – subsists in an intimate and yet agonistic relationship with mainstream nationalism. By exploring Gandhi's concept of swadeshi, I hope to point to how this politics, denying as it did the very necessity of a sovereign state, worked to question mainstream nationalism, and how it subverted the latter in the very process of sustaining it. Accompanying this neighbourliness was also a distinctive homelessness, one created by the discipline required for swadeshi.

Like Gandhi, Yagnik, in his more interesting moments, rejected mainstream nationalism. (My focus is only on such moments, since he made many other deeply problematic political moves. This article does not mean to reconcile various strands of his politics or evaluating them as a whole: indeed, this article is possible only because it sets aside the possibility of any such evaluation.) In his rejection, he drew on the problematic of a politics of neighbourliness. But this was a Gandhian politics directed against Gandhi himself. Gandhi, in his insistence on neighbourliness, had been unable to deal seriously or systematically with questions of marginality. His politics effectively evaded such questions by presuming that neighbourliness, practised seriously enough, would ensure that nobody was marginal. Yagnik, in contrast, insisted that the question of marginality could not be resolved this quickly. By bringing questions of marginality into systematic engagement with neighbourliness, Yagnik not only transformed the latter but also rendered himself homeless in the nation constituted by Gandhi's politics of neighbourliness.

I

The best place to begin is not the life of Yagnik (1892–1970), but that of the writer Kanaiyalal Munshi (1887–1971) – the figure with whom Desai had chosen to compare Yagnik.[2] Munshi was deeply involved with producing a distinctive love, simultaneously, for India and Gujarat. He saw his (widely read) historical novels as part of the process of regenerating India, remarking:

Bankim revived the memory of the heroism of the sanyas rebellion. Both he and Dwijendralal Roy resurrected Rajput heroism. Not that such authors intentionally set out to foster national or regional pride. They became the spontaneous voices of the coming spring. The same was perhaps the case with me.

(Sheth, 1979, p. 33f)

His novel *Gujarat no Nath* (The Lord of Gujarat) is quite characteristic of his concerns. First published in 1917, and translated into several Indian languages, it was set in twelfth-century Patan (Gujarat), and described confrontations among the Rajput kings of Avanti (Malwa), Patan, and Junagadh (Saurashtra) (Munshi, 1952).[3] A dominant theme in the book is the love for Gujarat. The policy of Munjal Mehta, the Prime Minister of Patan, had been 'very clear': it was to produce a love for Gujarat that rose above parochial concerns. In the novel, Gujarat itself is centred in what is now central and north Gujarat. The regions of Saurashtra, Kutch, and Lat (modern south Gujarat) are outside it; in other words, the task of constructing Gujarat remains. This is a task that is assigned by Munjal Mehta to Kak (the brahman warrior who is the central figure in the novel) as the book draws to a close: 'Remember, your aim is not to conquer Lal (modern south Gujarat) but to make it Gujarat' (Jotwani: 496). This, then, was an active love that performed the work of making Gujarat; such a love is only very inadequately described through the vapid constructivism involved in the invocation of imagined communities.

It was also a statist love. Though himself a Jain, Munjal had managed to keep religious disputes distant from the process of consolidating state power. (Formerly, the state had been debilitated by disputes between Jains and Shaivites.) 'His main objective was the growth of the kingdom and the establishment of a strong empire in Gujarat, and he considered philosophical and religious disputes utterly futile' (Jotwani: 81). This concern with producing a statist love for Gujarat recurs not only in many of his other novels but also in his social commentaries and histories, including *Gujaratni Asmita* (The identity of Gujarat), and *The Glory that was Gurjararashtra* (1954).

To understand this statist love better, a consideration of the tension between the love for Gujarat and another register of the same kind of love – one involved in making Bharat or India – is also important. Munjal, being a figure of the older generation, could not appreciate the need for this task of unity among the various states. For him, the transcendent love that rejected parochial identities stopped at Gujarat; it did not transcend Gujarat to include Bharat. The unity of Bharat that Munshi envisioned was, of course, of a particular kind, directed against the Muslim, or the 'ravaging Yavan hordes [that] are advancing steadily year by year'. The figure who articulated the need for unity against Muslims was young Kirtidev, an associate of the Avanti commander who had successfully laid siege to Patan and forced it to agree to a peace treaty. Pleading for a more extensive peace than the treaty would have provided, he argued: 'I pray to almighty god that the peace should prevail forever. Look, Patan and Avanti are like our country's two eyes. Doesn't it pain you when the two fight'. (Jotwani: 153) In other words, while Patan and Avanti may retain their identities, there should also be a non-antagonistic unity between them. To ensure a longer peace treaty, Kirtidev proposes that the Avanti king's daughter marry the king of Patan. However, Munjal, who despite his appreciation of Kirtidev's vision, fears that any alliance would give Malwa ascendancy over Patan, ensures that this plan is foiled. By refusing to spell out the consequences of this failure, Munshi produced a commonsensical knowledge among his readers: the 'knowledge' that it was the failure of Hindu kings to stay united which had led to Muslim rule in Bharat.

As this suggests, Munshi envisioned a particular kind of Bharat. Kirtidev appeals to Kak on the basis of his caste: '[Y]ou are a brahmin, captain. This country has been blessed and sanctified by you and your ancestors. If you do not come to her help, who will'?' (Jotwani: 204). This emphasis on the making of a Hindu Bharat was pervasive in Munshi's writings. Here, the emergence of the Indian nation is located in a historical past, specifically the Vedic and immediate post-Vedic period, which he situated in the first and second millennium BCE. Tracing the development of this already constituted subject, Bharat was for Munshi the task that Indian historians had to set themselves. One of the major projects undertaken by the Bharatiya Vidya Bhavan, the powerful educational institution that he founded in 1938, was the writing of a multi-volume history of India embodying precisely this spirit.[4]

However, I do not wish here to dwell on the exclusionary upper-caste, martial, and male Hinduism that constituted Munshi's Bharat. For now, I wish only to stress that this statist love – a love that authorized the subsumption of various regions within Gujarat, and of Gujarat within India – was constituted by a particular kind of emphasis on having a history. In the remarkably early 'Swadeshabhiman' (1856) – pride in/respect for one's own country – the poet Narmad, for instance, argued that while in India there was considerable *abhiman* or pride in the accomplishment of one's family or caste, an *abhiman* for the *desh* or country was not visible (Dave, 1975).[5] Nor was *swadeshabhiman* simply about having knowledge of or pride in the *desh's* past – that kind of pride already existed among Hindus, and the genealogical accounts of Bhats and Charans constituted that form of knowledge. Rather, it was about a particular kind of relationship with the past: it involved judging the past in moral terms (which the Charans did not do, hence the inadequacy of their histories) and building, in the present, on a moral pride in the past. This was what countries like Britain and France had done; it was what Hindustan needed to do. To acquire *swadeshabhiman* in this sense was crucially a matter of studying the past to establish the proper relationship with it. He had recently heard, he said, that a writer called Macaulay had written a two-volume history of England. In later essays in the book, he turned to writing this kind of history – providing a history of Gujarati literature and poetry, of his city Surat, of Gujarat, of the Rajput kings of Mewar, and of the *Ramayan* and *Mahabharat.*

As Narmad's argument suggests, to have a history was to love the country in a specific way: one where local affiliations like those involved in the family or caste were, rather than being rejected, subsumed in or transcended for a higher identity.

To have a history created through such transcendence was to create a stable object – be it Bharat or Hindustan or Gujarat – which could then be loved, and which could then shape the present. The nation-home had precisely the stability and resoluteness that Ramanlal Desai approved of in Munshi – it had a vision of India produced through history, and it worked to achieve this. It was this resoluteness that Narmad sought to foreground through his emphasis on *abhiman* or pride; indeed, for a later generation of Gujarati nationalists, *desh prem* or love of country was to become synonymous with *desh abhiman.* That very Nehruvian book, Khilnani's *The Idea of India,* recognizes and celebrates a similar resoluteness in Nehru: he, it states, 'rejected Jacobin notions of popular sovereignty... in favour of the idea of an abstract, historically durable "people" or nation'; in contrast, the book laments, later generations 'preferred to invoke the immediate, volatile authority of electoral majorities' (Khilnani, 1997, p. 41). This logic of transcendence, then, requires a stable object, usually created through history.

This logic of transcendence also produces the distinctive relationship between Gujarat and Bharat. Though both are constituted by the same logic – that of having a history – the relationship between them is one where Gujarat is subsumed under Bharat. Nationalist thought even had a name – regionalism (Munjal's crime) – to describe those moments when this hierarchy was not accepted: to be regional was to refuse transcendence, and to become parochial. Munshi was worried by the spectre of regionalism; thus his ambivalence toward the creation of the separate state of Gujarat. Munshi's own books were part of the early twentieth-century movement for such a state. In 1948, on the occasion of the establishment of Saurashtra as a state, Vallabhbhai Patel had suggested that all the states in Gujarat be merged under one administration. Later, Munshi organized a Mahagujarat conference, which again called for such a state. Yet, Munshi perhaps felt constrained by the prospect of losing the city of Bombay to Maharashtra, which he felt was inevitable if a state of Gujarat was formed. Also, he saw the movement for a separate state of Gujarat as potentially undermining the broader identity of India. Ironically, when the Mahagujarat agitation for such a state started, Munshi kept out of the struggle.

As the spectre of regionalism or volatile electoral majorities suggests, while the logic of transcendence sustained the nation-home, it also produced homelessness. The affirmation of a home is here undercut by its subsumption within a higher level of generality; regional loyalties and electoral majorities cannot really be a home. This was the curious sense in which Munshi was homeless within Gujarat: to affirm India was, within this logic, necessarily to not affirm Gujarat beyond a point. This kind of cosmopolitan homelessness was pervasive within the nationalist movement; it was the modality by which mainstream nationalists affirmed their loyalty to India. It was also the modality by which some of the most consistent nationalist thinkers affirmed a universal civilisation over the nation. Thus it was that Nehru sometimes felt stifled by the idea of being an Indian, or that Tagore insisted that nationalism was a destructive force because it was so parochial. Liberal cosmopolitanism not only produces the modern nation-home; it also produces the homelessness, perpetual exile, and even alienation of the citizen of the world who is also a citizen nowhere.

II

In Yagnik's writings, too, there is certainly a similar cosmopolitan logic which both creates various modern homes and simultaneously creates homelessness by subsuming these homes under higher levels of generality. As we saw, one kind of home that Yagnik claimed was among peasants and marginal groups, and indeed among the 'common people'. When he stayed in their midst in 1918, when he travelled in villages, his 'love for natural, outdoor beauty, for the farms and trees, which had remained suppressed in Bombay, was aroused'. And yet, this affirmation of peasants went hand in hand with a distancing from them. 'While my heart swelled with my love for the village, my intelligence and imagination were constrained. I was like the lover of an illiterate but beautiful woman' (Yagnik, 1971: 41).

The subsumption and even subordination of the popular to Gujarat are most evident in his account of the Mahagujarat movement, to which the sixth volume of his *Atmakatha* is dedicated.[6] As chairman of the Mahagujarat Parishad, Yagnik was perhaps the most important leader of the movement, which sought, in the 1950s, to carve out a separate state of Gujarat from the colonial administrative division, Bombay Presidency.[7] Partially as a result of the movement, Bombay Presidency was divided in 1960 into the states of Maharashtra and Gujarat. In Yagnik's account, the opposition to the Mahagujarat movement is stated as stemming only from a government and ruling party out of touch with the people. The identity of the people of Gujarat, in his account, is constituted by that long history of shared literature and culture; the movement is only about realizing this identity through a linguistically unified state. The invocation of the linguistic community of Gujaratis, then, glossed over the politically charged fissures present in the everyday use of the language, and over the relations of class, region, and gender that so differentiated the region. In the subsumption of the various regions within Gujarat, it practised a violence in no way different from that involved in Munshi's subsumption of Gujarat within India.

This is perhaps most strikingly evident in his account of the eventually successful struggle to make the Dangs region part of Gujarat rather than Maharashtra (Yagnik, 1971, pp. 346–348, 462–465, 478–479). The region was inhabited by forest communities, largely Bhils and Koknis. It clearly did not fit – in political, cultural, or linguistic terms – the schema envisioned in partitioning Bombay into two linguistically homogeneous states. Even more clearly than most other groups, the Dangis were neither Marathi nor Gujarati; most were little involved in the issue of linguistic states. Despite (or perhaps because of) this, as the struggle for the creation of Gujarat and Maharashtra heated up, the question of which state the Dangs should be part of became one of the most contentious issues. Yagnik devoted much energy to mobilizing the Gujarati moneylenders and timber merchants working in the region in order to ensure that the Dangs became part of Gujarat. There was a deep irony in this, since both were among the groups who exploited the Dangis. The creation of Gujarat thus involved the marginalization of the Bhils for whom his heart had swelled with love. Yet, it would be a mistake to treat this as a contradiction within Yagnik's practices or as a case of bad faith. Rather, precisely because the nation – whether Gujarat or Bharat – could, within this cosmopolitan logic, be made only by subsuming more particular identities, there were contexts in which moneylenders and merchants and *adivasis* had to join hands; the violence practised on *adivasis* was justified on the grounds that enabled them to affirm a less parochial identity.

Indeed, his *Atmakatha* itself was cast in part as an affirmation of Gujarat. As a shy person, he said, it was not that he wanted to write about himself, or that he thought his life was important. But he had been repeatedly asked to write his *Atmakatha* since he had been such a crucial actor in, and witness to, the making of modern Gujarat over the previous 50 years. (Yagnik, 1955, pp. 7–13) It is thus as a history of Gujarat that he casts his *Atmakatha* and himself one of the major makers of that history of which Gujarat would appear the natural subject and object.

Even Yagnik's marriage was subsumed within Gujarat and Bharat. For him, as for many other middle-class Indian nationalists, conjugal homes could only be created by companionate marriage.[8] And in his view, his own marriage was not companionate. By Yagnik's account, he had been engaged to his future wife when a child and did not wish to get married by the time he was in college at Bombay. He even wrote a letter breaking off the engagement. But under intense pressure from his mother, he gave in and married his betrothed, Kumud Tripathi. In the ensuing decades, he invoked his commitment to companionate marriage to practise a distinctively nationalist violence against Kumud. On the grounds that he had discovered that Kumud could not be educated to become the kind of nationalist-reformist wife he desired, he refused to stay with her. Finally, when he moved to Ahmedabad, he was persuaded to have her stay with him. But his refusal to have much to do with her drove her to attempt suicide. After this, her family in Nadiad took her away, and the two never stayed together again, though she repeatedly wrote to him requesting that they try being together again. She died in 1929, and Yagnik remained single.[9]

The mirror image of this violence toward his wife in the name of the nation was his concern for *stri-kelavni* (women's education). Here, as for many other nationalists, *kelavni* carried connotations of education, improvement, and training, directed not so much at personal advancement or neutral knowledge as at edifying the self and the nation. Among other things, *stri-kelavni* educated women for companionate marriages – in other words, for making nationalist homes. Such women – those with *kelavni – came* to symbolize the nation as a whole. It was these 'home goddesses' that he had met, Yagnik wrote, who came to mind when he tried to think of Gujarat. It was the households that they adorned, he said, which were his homes.

As this suggests, his celibacy did not always imply homelessness. It was also part of the embracing of another home – the nation-home. In choosing celibacy, Yagnik was acting within a long-standing nationalist tradition: recall, for instance, Aurobindo's famous call to youth to dedicate their lives to the nation. And less than two years after Yagnik chose celibacy for a second time, Hedgewar had founded the Rashtriya Swayamsevak Sangh (RSS), with its demand that all followers be celibate – that they be committed, in other words, to only one home. That celibacy was so often enjoined on nationalists is indicative of how the nation was seen to require the same kind of loyalty as the household did: both were homes organized on identical principles. When the committed male nationalist deployed the language of kinship, even though his form of address (where women became mothers or sisters, and men became brothers) was scarcely new, the logic on which it was based – that the nation was a household – home organized on the principle of identity – was quite novel. It was as if as a male that Yagnik had access to the households of other middle-class Gujarati nationalists. And precisely because he had access to these households as a figure committed exclusively to the nation-home, he felt that developing a romantic relationship would have violated the sanctity of the nation-home, and that, therefore, celibacy was almost enjoined upon him. He realized, he writes, that if he did not take a vow of celibacy, all the places that were open to him as refuge would be closed. 'I would be obliged to fly away from them'. (Yagnik, 1956, p. 265).[10]

Gujarat, in turn, was subsumed under Bharat. From around 1907, the

> ideals of social service and national freedom continued to grow in my mind. Hind maiya [Mother India] became the highest Goddess to be worshipped, the whole of Bharat a vast temple of Goddess; now the uppermost goal for me became service to its 33 crores children, and the struggle for their freedom.
>
> (Yagnik, 1955, p. 130)

Yagnik insisted repeatedly that the Mahagujarat movement was deeply committed to – and finally subsumed under – the idea of Bharat: he claimed legitimacy for the movement by invoking major nationalist leaders from Gujarat like Gandhi and Vallabhbhai Patel. And he felt homeless at times even within Bharat. This was especially so in the 1920s, when he insisted that India had a lot to learn from Western civilisation, but the theme was present in later years too.

III

However, this cosmopolitanism was, in Yagnik's case, always accompanied and undermined by another modality of being homeless: that produced by the marginality of a politics that affirmed the peasant, that which questioned the logic of transcendence. Consider, for a start, his ambivalent relationship with the Congress. In 1917 and 1919, he had been involved in providing relief to the Bhils and other communities in the Panchmahals, which was reeling under the effects of a famine. By 1921, he felt that these relief works should be made permanent, and in 1921 he and Amritlal V. Thakkar (better known by his later name of Thakkar Bappa) went ahead and started a Rashtriya Bhil Ashram at Mirakhedi village. Yagnik also started a school for untouchables, and, to meet the expenses, made a budget for Rs 5,000 and submitted it to the Congress Provincial Committee. Vallabhbhai, however, refused to approve this: while the Congress was committed to a campaign against untouchability, he said, it had to direct its financial resources to the struggle against the British. Yagnik's arguments invoked marginal groups:

[H]ow long can we watch as the defenceless and half-naked Bhils are looted by government and *sahukars,* and are hurled into starvation with every failure of the rains?... [D]oes swaraj mean anything for them?... If our fight for *swaraj* is not for a fistful of bhadra people but for the *daridranayaran* [poorest of the poor, God in the poor] then ... we should establish ideal institutions to demonstrate our unity with those backward people, and to show them the governance of the future.

(Yagnik, 1956, p. 14)[11]

When Yagnik's arguments had little effect with the committee, he approached Gandhi directly and secured approval. This upset Vallabhbhai, who remarked: 'How long will you go over our heads and get your work done through Gandhiji?' (Yagnik, 1956, p. 21). Feeling that there was little point in staying in the Congress in the face of such hostility from many senior members, Yagnik submitted his resignation from the Congress, which Gandhi promptly accepted.

Note how the trope of the 'common people' figures in the debate. Vallabhbhai and others within the Congress did not question the centrality of the people, or of working for them. But theirs was the abstract, historically durable people constituted by having a history. Their arguments were the traditional ones made by mainstream nationalists: that the money did not exist for the purpose, that *swaraj* was for the moment the more important goal, and that projects such as this should ideally be undertaken after *swaraj*. Indeed, Vallabhbhai saw himself as committed to the people, yet Yagnik's resignation was, for him, about the refusal to accept the discipline needed to serve the people *by* serving the nation:

To resign, and to free oneself – it is more painful than that to remain inside without resigning. Indulal is my younger brother. We have lived together as brothers till today, and now matters have come to this. What can I say. Words fail me. I cannot speak further.

Saying this, the brave Patidar leader, 'the *naik* [leader] of Gujarat, sat down, trying to hold back the tears in his eyes'. (Yagnik, 1956, p. 25) The nation that Patel invokes, in whose name he defers the setting up of *ashrams* for Bhils and untouchables, is, of course, the nation constituted through the claim that it had a history.

Similarly, though marginal groups are virtually never actors or a significant presence in their writings, Munshi and Nehru claim to speak for them by subsuming them in the imaginary of the popular. This imaginary of the popular is inseparable from the logic of transcendence associated with nationalist history. It is in this sense that nationalist history was part of a mediatory project: it claimed to be Indian modernity, and to thus know and subsume the popular within it. The task that nationalist thought set itself, therefore, was that of mediating between the popular and the modern, of making the popular into the modern nation or, more precisely, in the image of the nation already provided in and by history. By situating Indian identity not in the consequences of contemporary politics involving peasants or other marginal groups but in the imaginary of the popular, the implication of this politics could be circumscribed: it was legitimate to the extent that it affirmed a people – Khilnani's 'abstract, historically durable "people"' – already provided through history.

Yagnik, in contrast, was sceptical of this understanding of 'common people' through history. He felt that his fellow Congress members 'continued to look at the rural people of Gujarat from an urban point of view'. As for him, 'the conviction that my identity with this different rural world, with its miserable and oppressed people...would become permanent – this conviction grew within me' (Yagnik, 1956, p. 26). Here, then, the people come to have a different connotation: rather than having the solidity given by having a history, they are characterized by a marginality. And this is a marginality that explicitly challenges the notion of the popular constituted through history. Thus, not only are he and Vallabhbhai different from the marginal 'rural people,' but it is the latter, in their marginality, who constitute the nation.[12]

In the 1930s, around the time that he had just started publishing a periodical called *Khedut Patrika* (Peasant Newsletter), his own sense of identity with the *kisans* was almost literal.

> I would remember the thousands of naked hungry Bhils that I had seen at Meerakhedi. I would imagine their incarnation as one immense, powerful, world-father kisan, and, becoming one with them, I assumed the form of that huge kisan, with my head reaching the sky and my feet going deep down in the soil. In this mood, I would take long strides to my host's place and draw everybody into stories of kisans. In this fashion, turning into a proper kisan, I got ready to work for the kisans.
>
> (Yagnik: 43)

True, there was often a seeming convergence between the *kisan* and the nationalist trope of the popular. Consider how Yagnik characterises the rule of Vanraj Chavda, one of the Hindu rulers of Gujarat. Speaking at a school, he characterised Vanraj Chavda as a *gurjarvir,* or a warrior of Gujarat, rather than simply a king, because he said, despite being a king, Vanraj Chavda had lived with the Bhils; Champa Vaniya (a prominent merchant) had helped him so much that Vanraj named the fort of Champaner after him; the Jain *muni* Shilgunisuri (an important priest) had instructed him; and the Bharvad Anhil (head of the powerful Bharvad community) had helped him so much that Anhilwad Patan was named after him. Thus, Yagnik argued, it was through the support of a range of groups that Vanraj Chavda established rule over Gujarat (Yagnik: 13).

But in Yagnik's case, the argument is not that Vanraj is a people's king – that he represents the popular. It is a subtly different one: it insists that there is a distance between the ruled and the king, that the forging of a relationship across this distance cannot be taken for granted. It is after Vanraj forges such a relationship that he becomes a king. Thus, in contrast to the claim to have a history – which constitutes the nation as an already known entity – this account rests on the inadequacy of Vanraj, the absence at the centre of the nation. This absence has to be filled by an engagement with various groups, but the question of whether it has been filled can never quite be settled. The presence of *kisans* and rural people does quite a different work from the imaginary of the popular – by emphasizing the need for active affirmation from marginal groups, it introduces a fundamental instability in the nation. To ally with 'rural people' in this sense was indeed to become an *asthir man na fakir*.

The prominence of figures like Yagnik – and they are legion through the twentieth century – may suggest a need to revisit our usual understandings of Indian politics and nationalism. We have often been misled by the prominence of figures like Bankim, or Nehru for the later period, into presuming the hegemony of a cosmopolitan historical vision. We presume that the mainstream nationalist discourses of history, citizenship, or secularism, with all their exclusions, have been constitutive of the modern Indian state. And it is certainly true that both the Nehruvian Congress and the Bharatiya Janata Party (BJP) have shared the emphasis on having a history, though, of course, in very different ways. This is also why many of us have sought to criticise precisely these discourses. Through our criticisms, we seek to criticise the modern Indian state, to argue that the historical, the secular, and the citizenly are not simply means of liberation or empowerment but are also forms of domination.

Still, while these discourses are a crucial part of modernity's self-representation, surely it is in the failure of this cosmopolitan vision to hegemonize the polity that an understanding has to be sought for many of the forms of everyday politics in colonial and postcolonial India. For without a fairly restrictive and pre-Foucauldian understanding of the state, it would be difficult to sustain the argument that a Nehru-Patel-Munshi vision—which shared an emphasis, despite differences, on secularism, history, and citizenship – constituted the state. True, it did constitute what could, in an unhappy phrase, be described as the formal apparatus of the state. But what of those myriad spaces of government – and, even more, of politics – outside that state, where other practices of power, no less modern if less symbolic of the modern, were influential? These practices are what we sometimes describe, suggestively but inadequately, as plebiscitary and populist politics; these forms are what occasion the laments of those like Khilnani who are committed to the historical vision that is more resolutely cosmopolitan. It is in understanding the politics involved here – the politics of the ahistorical, asecular, and acitizenly (neither pre-, nor anti-) – that figures like Gandhi, Yagnik, and myriad others who had a conflicted relationship with mainstream nationalism are of particular significance. Indeed, tracing and accentuating the tension of such politics with the mainstream nationalism that remains dominant is perhaps one of our most pressing tasks as subaltern historians.

IV

How do we conceptualize this politics – the politics that rejected the logic of transcendence and affirmed instead the ahistorical, asecular, and acitizenly? In his rejection of the logic of transcendence that constituted mainstream nationalism, Yagnik worked his way through the thought and politics of Gandhi, whose own questioning of mainstream nationalism was at the time particularly influential not only in Gujarat but also in the Indian anti-colonial movement more broadly. The centrality of *ashrams* in Gandhi's politics is symptomatic of such questioning. As we know, he stayed more or less continuously in *ashrams* or in *ashram-like* institutions from around 1904: the Phoenix settlement and Tolstoy farm in South Africa; the Satyagraha *ashram* in Ahmedabad (1915–1931), and Wardha and Sevagram *ashrams* in later years. Mainstream nationalists such as Nehru were often frustrated about the amount of time Gandhi spent in and on the tiny institution of the *ashram* – *they* regarded this as a wasteful eccentricity in a man who was after all the principal leader of the nationalist movement. Gandhi, obviously, did not feel this way. He said of the Satyagraha *ashram* that it 'set out to remedy what it thought were defects in our national life'. He meant this in complete seriousness: as I have argued at length elsewhere, if *Hind Swaraj* represented Gandhi's critique of liberal modernity, then the *ashram* was the conceptual site through which he tried to think about and develop practices for an alternative politics that questioned the nation constituted by the logic of transcendence. This politics is most explicitly spelt out in two books which deal with the vows or observances (as he interchangeably translated the Gujarati words *vrat* and *yama)* involved in *ashram* life. The first, *Mangal Prabhat* (Tuesday Dawn/Auspicious Dawn, translated into English as *From Yeravada Mandir)* focused on the vows involved in *ashram* life, titling each chapter after the vow it discussed. The second, *Satyagrahaashramno Itihas* (A History of the Satyagraha Ashram, translated into English as *Ashram Observances in Action),* shared many of the same chapter headings, but now dealt more directly with what these vows meant for everyday life in the *ashram.*[13]

Between them, these (approximately) 11 vows – the precise number kept changing, but they usually included the vows of truth, ahimsa or love, celibacy or *brahmacharya,* control of the palate, poverty, swadeshi, fearlessness, and removal of untouchability – articulated a very precise alternative politics. I have elsewhere explored the politics involved in many of these concepts. Here, I shall limit myself to focusing in particular on the vow of swadeshi (literally, of one's own *desh,* or country), where his distance from the logic of transcendence is perhaps articulated most forcefully. As we know, swadeshi was a concept which was already in place before Gandhian politics became influential in India – a swadeshi which involved the use only of locally made goods, and the boycott of British goods, was practised during the 1905 agitation against the partition of Bengal. Gandhi, too, repeatedly called for swadeshi and urged, in particular, that everybody take a vow to wear only swadeshi clothes (he felt that a swadeshi vow that was inclusive of other items would be too ambitious). But his was a distinctive concept of swadeshi, as different from the earlier swadeshi as his *ashram* was from those it acknowledged as its precursors. Considering the matter in his characteristically careful fashion, he made the concept of swadeshi an occasion for specifying what was local or native, what was foreign, and what was universal.

To use foreign articles rejecting those produced or manufactured in India is to be untrue to India, it is an unwarranted indulgence. To use foreign articles because we do not like indigenous ones is to be a foreigner. It is obvious that we cannot reject indigenous articles even as we cannot reject the native air and the native soil because they are inferior to foreign air and soil.

(Collected Works, 1919, p. 43)

This invocation of native air and soil may, on a quick reading, seem redolent of that nationalism associated with Herder or Mazzini, which treats the nation as a natural fact constituted by blood, soil, race, or language. Such nationalism has been questioned in our times by the emphasis on an imagined community. Yet, despite rendering as contingent constructions that which the former had presented as natural facts, this shift in emphasis operates with the same elements. In contrast, Gandhi's invocation of native air and soil involved a radically different move.

Here, the native had nothing to do with a shared culture, history, or even political experiences. Thus, swadeshi never seemed to refer to any finite, clearly mapped boundaries. Rather, swadeshi was that which was insistently local in a way that turned its back on transcendence. 'The purest swadeshi vow will be to use cloth made out of yarn spun by one's wife, sisters and children in the home' (CWMG, 1919, p. 117). Similarly, while addressing a group of women at Nadiad, Gandhi emphasized that the focus should be on weaving *khadi* cloth primarily to meet Nadiad's own needs, and then that of surrounding areas.

Swadeshi thus centred around the neighbour or *padoshi,* a concept that could not be identified with a nation produced by history, culture, or the people; a concept that had very clear obligations but no clear boundaries. In 1915, when he had made swadeshi one of the vows of *ashram* life, Gandhi invoked the logic of neighbourliness to explain swadeshi:

Man is not omnipotent. He therefore serves the world best by first serving his neighbour. This is Swadeshi, a principle which is broken when one professes to serve those who are more remote in preference to those who are near.

(CWMG, 1928, p. 109)

Here, then, love of country was constituted through *seva* (or service) to and *prem* (or love) of the neighbour rather than pride in history, culture, or people.

The neighbour or *padoshi* is a crucial concept in Gandhi's writings. Indeed, neighbourliness might be one of the best possible renderings into English of *ahimsa,* one of Gandhi's most important vows. In the English translations of his writings (which Gandhi himself usually supervised or revised, and which he occasionally undertook himself), Gandhi avoided translating *ahimsa* as 'non violence'; his preferred translation was 'love'. Thus, for instance, the chapter on *ahimsa* in the English versions of *Satyagrahashramno ltihas* and *Mangal Prabhat* is titled, in both cases, '*ahimsa* or love''; in the Gujarati version, the title is simply '*ahimsa.* Gandhi's addition of the word 'love' changed the connotations of the English equivalents. *Ahimsa,* when translated simply as non-violence, was quite congruent with the liberal notion of civil society (which is constituted, after all, by the absence of violence – hence its civility); the emphasis on love, rather than the neutrality or negativity of non-violence, foregrounded Gandhi's distance from liberalism. That emphasis on love, of course, has created its own problems (especially because scholarship has been so surprisingly dependent on the English translations of his writings rather than using these along with his Gujarati writings): Gandhi's politics has often been understood as some form of spiritual universalism. To understand *ahimsa* as neighbourliness, instead, entirely avoids this connotation of spiritual universalism and is in keeping with the spirit of Gandhi's rendering of *ahimsa.* The concept of the *padoshi* or neighbour was a familiar one in Gujarat. Babu Suthar has pointed out that *'padoshi dharma'* (it is possible that when Gandhi used the English phrase 'law of love' – and in most places, his Gujarati originals do not seem to have a parallel for this phrase – he was referring to this) is a very common phrase in Gujarat; it has usually referred to the moral principles which guide relationships with neighbours. As a *padoshi,* one had certain claims on neighbours, and hence the common phrase *'padoshi pahelo'* or neighbours first.[14]

I have explored elsewhere how this neighbourliness led, for Gandhi, to two ways of relating to the neighbour – friendship and service.[15] For now, suffice it to note three implications of this emphasis on neighbourliness that are particularly salient to understanding the distance and tension between it and the logic of transcendence. First, there was the question of the relationship to that which was being rejected – the foreign. Because the foreigner, too, was a neighbour, even if a more distant one, Gandhi insisted on the difference between swadeshi and boycott. A boycott of British goods and cloth would have been 'a purely worldly and political weapon [in Gujarati, *rajya-prakarni,* or statist]...rooted in ill-will and the desire for punishment'. In contrast, swadeshi was 'the natural duty imposed upon every man' (CWMG, 1919, p. 396). He sought a 'swadeshi in a religious and true spirit without even a suspicion of boycott' (CWMG, 1919, p. 184f), a swadeshi in which even the British Viceroy would be able to take part. Put differently, the logic of boycott was that of mainstream nationalism. While it contested the imperialist argument that the interests of the colony had to be subsumed to those of the empire, it did so because of its claim that the Indian nation was the true culmination of the logic of transcendence. As such a culmination, the relationship of the Indian nation to that which was foreign was necessarily one of antagonism, hence the possibility of a boycott. Because Gandhi's 'native' was constituted by a neighbourliness rather than transcendence, it did not have this relationship of antagonism with that which lay outside it. Thus, the vow of swadeshi did not seek, necessarily, to end British rule, but rather to transform it: if it were followed, 'even British rule will cease to be foreign rule and will become swadeshi rule' (CWMG, 1919, p. 116).

Second, there was Gandhi's rendering of the universal. In 1930, revisiting the question of swadeshi as an *ashram* vow, he again opposed this principle of serving one's immediate neighbour to that of serving the world in more general terms.

On the other hand, a man who allows himself to be lured by 'the distant scene' and runs to the ends of the earth is not only foiled in his ambition but also fails in his duty towards his neighbours. Take a concrete instance. In the particular place where I live, I have certain persons as my neighbours, some relations and dependants. Naturally, they all feel, as they have a right to, that they have a claim on me, and look to me for help and support. Suppose now I leave them all at once, and set out to serve people in a distant place. My decision would throw my little world of neighbours and dependents out of gear, while my gratuitous knight errantry would, more likely than not, disturb the atmosphere in the new place. Thus, a culpable neglect of my immediate neighbours, and an unintended disservice to the people whom I wish to serve, would be the first fruits of my violation of the principles of Swadeshi.

(1932, p. 62f)

Running through these arguments is a concern with how neighbourliness allows one to 'serve the world'. But rather than seeing a tension between the two, he insisted that pure service of the neighbour was also service of the world.

At the Ashram we hold that Swadeshi is a universal law. A man's first duty is to his neighbour. This does not imply hatred for the foreigner or partiality for the fellow-countryman. Our capacity for service has obvious limits. We can serve even our neighbour with some difficulty. If everyone of us duly performed his duty to his neighbour, no one in the world who needed assistance would be left unattended. Therefore one who serves his neighbour serves all the world. To serve one's neighbour is to serve the world. Indeed it is the only way open to us of serving the world.

(*Satyagrahaashramno Itihas*: 213; CWMG, p. 172)

Indeed, he argued that the 'votary of swadeshi' should become indistinguishable from those who live with us through service to them. This may make it appear as if there could be exclusion and even sacrifice of the rest. But it is not so. Pure service of our neighbours is also necessarily service to the foreigner. 'As with the individual [from *pind*], so with the universe [*brahmand*]'.[16] Involved here is a radical reworking of the concerns that characterized the logic of transcendence. Like the latter, swadeshi, too, claims to provide a 'universal law'. But this universal law, unlike the latter, cannot be arrived at by subsuming the particular under ever higher levels of generality. The concept of the neighbour is constituted by singularity rather than particularity, and it is only through service of this singular that the universal can be reached.[17]

Third, there was the relationship of swadesi with sovereignty. In mainstream nationalist understanding, to become a nation-state was to become a sovereign community, whether natural or imagined. This sovereignty, in its nationalist form, involved a distinctive relationship with the logic of transcendence. While the logic of transcendence provided the norm by which the nation-state constituted itself, the latter's sovereignty also had to be constituted by the latter's sovereign denial of that very norm. On the one hand, thus, sovereignty over the components of the nation – its various localities – was claimed precisely on the basis of transcending them.

And yet, to claim sovereignty was also to break with this logic; it was to insist that there could be no further subsumption that the nation-state was sovereign. It was this very claim to sovereignty that authorized the nation-state to impose a logic of transcendence on its various localities, and that also authorized the nationalist paradigm of development, where violence was practiced against people in the name of the people.

In contrast, neighbourliness did not, as a concept, allow for the sovereign nation-state. Gandhi's hostility to such a state is, of course, well known. As early as *Hind Swaraj,* he had remarked to his liberal nationalist interlocutor who wanted such a state:

> [You] want English rule without the Englishman. You want the tiger's nature, but not the tiger; that is to say, you would make India English. And when it becomes English, it will be called not Hindustan but Englistan. This is not the Swaraj that I want.
>
> (CWMG, p. 255)

For him, especially by and after the mid-1920s, freedom from British rule was sought not in order to establish a post-independence nation-state to replace the British state, but rather because the British state hindered the pursuit of swaraj, as he feared that even an Indian nation-state committed to 'modern civilisation' might. This is the sense in which Gandhi's homelessness was radically different from that conceivable from within a cosmopolitan vision.

As these implications of swadeshi indicate, Gandhi's homelessness was not the cosmopolitan one possible within the terms of the mainstream nationalist problematic; it was outside the very problematic that produced the nationalist home and homelessness. At the same time, the practices of the *ashram* and of neighbourliness produced their own home, but they used a very different set of procedures for thinking of home and homelessness. Indeed, Gandhi was on several occasions to refer to Gujarat and India as his home, and in doing so, he was deploying the politics of neighbourliness to constitute the home. For him, to make a place a home was to be neighbourly; any other way of claiming a home was to create 'Englishtan'.

This home had its own discipline, just as the nationalist home had a discipline where the particular was subordinated to the general (and it is this discipline which authorizes development as a nationalist project) (Skaria, 2003). The *ashram* was the principal site for this discipline that produced neighbourliness. It is because of the centrality of discipline in creating the nation as a neighbourly home that Gandhi was never quite able to describe swaraj (self-rule) as a birthright or any sort of right; he insisted famously, as we know, on the duties and restraint involved in swaraj. Indeed, as early as 1908, he insisted on the centrality of neighbourliness in the making of the nation: arguing that America, France, or England did not have 'real swarajya', which was enjoyed only by the man who 'does his duty' to his family, 'his servant, and his neighbour'. This swaraj was quite independent of and even prior to the creation of an autonomous nation-state. 'Such a man will enjoy swarajya wherever he may happen to live' (CWMG, 1908, p. 458).

Of course, as with the logic of transcendence, the neighbour, too, needed to be transformed; despite his celebration of villages, he remarked that they needed to 'shed their laziness and make a corporate effort to live' (CWMG, p. 29). But unlike the discipline involved in development, the discipline involved in bringing about this transformation was directed entirely at the self. This was so not in the severe nationalist sense familiar to us from the celibacy of the RSS, where self-discipline authorized the disciplining of the neighbour. In Gandhi's argument, nothing could authorize the direct disciplining of the neighbour. Rather, it was through a discipline of the self that a non-coercive conversation with the neighbour could begin. Thus, while he described 'the Bhils, the Pindaris, the Assamese, and the Thugs' as 'jungli,' he insisted that they are 'our own countrymen' and that effort had to be made to 'win them over' (Gandhi, 1997, p. 45). Here, then, the tribes and villagers in Gandhian discipline were not converted into backward groups requiring that historicist staple – development; here, there was no conceptual space for Vallabhbhai 's tearful subsumption of the peasants within the nation's long-term interests. It was in principle (though a principle more often observed in the breach in the various *ashrams* set up among such groups by Gandhi's followers), a mutual conversation, one in which the interlocutor could also transform the Gandhian. His support for figures like Nehru or Patel is best understood in these terms: neither as a contradiction that reveals the true class character of his politics nor as an anomaly that should be ignored, but rather as an attempt to transform the dominant logic of transcendence through the practice of neighbourliness, to supplement the nationalist home with the neighbourly home.

The discipline that produced neighbourliness also created its own homelessness. 'Ashram here means a community living religiously together. As soon as I set up house, my home was like an ashram in two senses, for *grahastha-ashram* is not about pleasure [bhog] but duty [dharma]'. (*Satyagrahaashramno Itihas*: 186; CWMG, p. 142). Because neighbourliness itself was constituted by duty, there was no place in it for the kind of attachment – Gandhi's 'pleasure' – and immediate personal ties that usually constituted home. This was the sense in which true neighbourliness was also simultaneously about a homelessness. Gandhi admired this kind of homelessness. Clarifying a misunderstanding once, he wrote:

I paid you a compliment by summing up your life as of a homeless wanderer. I connected you with *aniketa* [the state of being without a home] of the *Gita* and envied you. Your home was nowhere and everywhere. How could you mistake all this for a reflection on you. It shows what a sorry thing foreign speech is.

(CWMG, 1944, p. 35)

The practice of this kind of *aniketa* led, strikingly, to the *niketan*. For many early twentieth-century Indian thinkers, the *niketan* was the home constituted by the practice of spiritual discipline: thus Shantiniketan and Anandniketan, amongst others. But the spiritual discipline of the *niketan* was often identified with the logic of transcendence (as with the Arya Samaji *ashrams),* and what Gandhi attempted was to constitute the *niketan* instead through a politics of neighbourliness. A different home and homelessness, then, existed together within the politics of neighbourliness.

V

Yagnik's homelessness was produced through an engagement with Gandhi's politics of neighbourliness. He was, of course, a close associate of Gandhi. He had been associated with Gandhi almost since the latter's arrival in Gujarat from South Africa. Indeed, Gandhi's main Gujarati weekly, *Navjivan,* had been initially started by Yagnik in 1915, quite independently of Gandhi. Later, at Gandhi's request, the weekly was effectively turned over to him, with Yagnik doing most of the editorial work in the early years. Yagnik was also in Yeravada jail with Gandhi; during this time, he took extensive dictation from Gandhi.

In the early years, when a staunch follower of Gandhi, Yagnik played a crucial role in setting up *ashrams* in the Panchmahals and Kheda. When he set up an *ashram* for untouchables in the Panchmahals, he secured a building away from their locality, 'for only then could we impart sound cultural influences to their children'. Working with the Bariya-Kshatriya community of the region, he remarked that they had been given a bad name and harassed on charges of drinking and theft. It was not enough to provide them with the spinning wheel and with ideas of prohibition; there was rather a need to 'colour them with new cultural influences like other backward communities'.[18] Yagnik's early *ashrams* thus performed tasks quite congruent with those that Gandhi emphasized: they provided discipline, especially important in their attempt to reform Bhils and untouchables *kelavni.* It is symptomatic of Yagnik's closeness to the Gandhian paradigm that he was to write, while imprisoned in Yeravada, that if nothing else in the world suited him, then he would go to the Mirakhedi Bhil *ashram,* which would always welcome the 'homeless traveller' and spread his mattress there.[19]

By the 1940s, however, when Yagnik set up his *ashram* at Nenpur near Mehemedabad, just a short train ride away from Ahmedabad (this is the *ashram* most closely associated with him), it was quite unlike the Gandhian *ashram.* Though he remained single, the conventional Gandhian emphasis on discipline disappeared. There was *kelavni* here too, not only through schools but also agricultural training. Nevertheless, the position that it occupied was different. Thus, in his accounts, by this time, the emphasis was less on how this *kelavni* would transform the 'harijans' and more on the confrontations that this created with the *ashram's* neighbours-upper caste orthodox Hindus. On learning that not only were there harijans at the *ashram,* but that they ate and drank with other village students, Bansiwala Maharaj, the powerful upper-caste religious leader in the area who had permitted Yagnik to use the land for the *ashram,* led a large procession of villagers to the *ashram* and demanded that they vacate the spot immediately. Yagnik consented by moving to another site, but threatened legal steps. Indeed, the tensions between the upper castes and the *ashram* remained high, with Yagnik making few conciliatory moves. Yagnik's actions, in this case, were early indicators of the growing transformation of his *ashram* primarily into a site for politically organizing peasants and workers.

Involved here was not only a rejection of the logic of transcendence (to organize peasants and workers for a radical politics was, effectively, to question the independent Indian nation-state that was constituted by this logic) but also, more interestingly, a questioning and radicalization of Gandhi's politics of neighbourliness. Two lines of questioning are particularly salient here. Gandhi's emphasis on service and love of the neighbour was not attentive to the political relations that constituted the neighbour. True, one of the vows of the *ashram* was for the removal of untouchability, and Gandhi campaigned ceaselessly for it. Similarly, Gandhi sought to serve the *daridranarayan,* as we saw Yagnik reminding Congress members. But neighbourliness as a form of love did not systematically take account of such matters. In this sense, there was a serious tension and, perhaps, contradiction between the Gandhian vow for the removal of untouchability – which broached the question of the marginal – and that of swadeshi. The agonized playing out of this tension in Gandhi's writings and politics is so complex that it needs to be addressed separately rather than in passing here. For now, suffice to note the first question that Yagnik's politics effectively posed: what when the neighbour to be loved and served was constituted not simply by singularity but by singularity *and* marginality?

Even in his early days, Yagnik had been sceptical of the Gandhian deployment of the concept of neighbourliness:

I had written to Thakkarbapa even from jail that if nothing would suit me in the world, finally, I would spread my mattress in the Bhil ashram of Meerakhedi. Still, within a short while of going to such places and seeing the oppression by government departments of the poor peasants, a fire would flare within me, and I would become impatient to fight it and awaken the peasants. But usually in such ashrams, there was a tradition of keeping one's head down and looking after the needs of students, concentrating only on education, without confronting the government. This I could not accept. Without some vision about the future of the entire peasant community, without doing some daily activity, I could not be part of such an institution.[20]

The love involved in neighbourliness, in other words, did not seem to deal seriously enough with the antagonisms that constituted the everyday lives of marginal groups.

This reservation above the love or *prem* involved in Gandhi's neighbourliness was one that Yagnik returned to repeatedly. In the early 1920s when Gandhi was upset about the violence in the national-level satyagraha movement, Yagnik insisted that love was not the criterion by which to judge 'ordinary people'; rather their anger should be regarded as natural and just (Yagnik). Later, he wrote an article attacking Gandhi's focus on love. Often, these reservations were cast in the language of transcendence; thus, his attack in the article was partially on the grounds that the latter did not allow for a scientific worldview. Nevertheless, his politics (always far more sophisticated than his analyses and writings) through most of this period drew heavily on the trope of love. And this was a particular kind of love for the 'ordinary people', where ordinariness signalled marginality. To love ordinary people in this sense, however, turned out in Yagnik's politics to be forcefully confrontational of other neighbours. This unresolvable antinomy between the politics of neighbourliness and marginality may perhaps also help us understand the profound irresoluteness of Yagnik's politics.

The second question that Yagnik's politics effectively posed concerned the self that was the object of self-discipline. Yagnik had been critical of the *ashram's* self-discipline from quite early on. After 1918, as the editor of *Navjivan,* he used to go to Sabarmati *ashram* regularly to have proofs corrected and sometimes stayed overnight. But the daily routine of the *ashram* chafed on him – his taste for reading late at night could not be fulfilled since the family with whom he stayed at the *ashram* rose at 4.00 am. Also, he was in the habit of drinking several cups of tea in the morning, and this was not possible. Finally, he gave up on staying in the *ashram*; he would stay in the city overnight and would cycle over to meet Gandhi in the morning (Yagnik). Some of this irritation surely was and remained nationalist – a sense that such discipline was irrelevant to the task of securing Indian independence.

However, in its more interesting moments (and I do not wish to claim that these moments were dominant in his politics), the question that was involved in his irritation was also this: what when the emphasis on loving the neighbour blurred the distinction between the self and the neighbour, not in the transcendent sense where the self subsumed the neighbour, but rather in the sense that the self itself becomes fragmented or suffused by the neighbour? (Though a serious consideration of this issue will have to await another occasion, it should be noted here that, in its simultaneous emphasis on affinity and difference, Gandhi's concept of the neighbour covered, in a strikingly different manner, some of the same territory as that covered in European philosophical thought by the concept of the other.) Again, the possibility for this question was created by Gandhi's politics of neighbourliness, but his emphasis on self-discipline defused many of its radical implications. Two implications of the austerities of self-discipline are particularly relevant here. First, these austerities were meant to make possible conversation with the *daridranarayan – the* neighbour who was most common in India – by becoming more like the *daridranarayan* to be served. Second, and this has not yet been emphasized enough in this article, the austerities were also meant to create the physical space for the neighbour – and here the neighbour included all living things – by reducing wants. Thus, Gandhi insisted:

If I save the food I eat or the clothes I wear or the space I occupy, it is obvious that these can clearly be used by those poor whose need is greater than mine. Since my selfishness prevents him from using these things, my pleasure [bhog] involves violence to my poorer neighbour. When I eat cereals and vegetables in order to support life, that means violence done to vegetable life.
(*Satyagrahaashramno Itihas*: 202; CWMG, p. 160)

Some violence was thus unavoidable for the sustenance of human life, but to reduce this violence through self-discipline created more space for others, Gandhi insisted.

Nevertheless, this communication and space were produced by insisting on the integrity of the self; it was this integrity that made self-discipline possible. In contrast, though Yagnik's politics sought the same conversation, the intensity of his engagement with the *kisan* often (though obviously not always, as his involvement with the Mahagujarat agitation indicates) rendered uncertain the distinction between him and the *kisan*. In this sense, self-discipline as a way of communicating with the neighbour became less important. And yet, not quite. For, once again, this tension between self-discipline and a love for the neighbour that went so far as to render uncertain the distinction from the neighbour was an antimonic one. Self-discipline was important, within the politics of neighbourliness, for creating, if nothing else, the physical space for the neighbour. To reject self-discipline without rejecting the politics of neighbourliness: this was the concern that drove Yagnik's politics in its most interesting moments. It was a concern that he never quite managed to address adequately, and the consequent shuttling between self-discipline and its rejection was yet another irresoluteness that marked Yagnik's politics.

Yagnik's questioning of self-discipline, despite its almost necessary irresoluteness, had one distinct consequence. It was self-discipline that constituted both home and homelessness within the politics of neighbourliness; to reject self-discipline was to refuse the *ashram* as a home as well as to refuse the detachment of *aniketa* as homelessness. In this sense, Yagnik's homelessness was not only outside the nationalist problematic (as was Gandhi's); it was also almost inconceivable within the problematic of the politics of neighbourliness.

Homeless in the nation and homeless in the *ashram:* why did this homelessness not lead to an affirmation of the figure of the subaltern and the marginal as a home – to a home, say, in an affirmation of the *daridranarayan,* or of *kisan* or *khedut* lifestyles? Such a home, we know, is claimed by many forms of radical political romanticism; it is the desire for and claim to such a home that has produced some of the most thought-provoking of our radical ethnographies and histories. Yet, Yagnik never quite claimed such a home. He was heavily involved with the everyday lives of the peasants, and the *Atmakatha* does provide accounts of these struggles. But there is little that is written about the *kisans'* everyday lives, save in the context of the oppression they faced.

Why should this have been so? We would certainly not be incorrect in treating it as a symptom of his nationalist or Gandhian affiliations: perhaps he claimed no home among them because of his distance from them, because he was not interested in them save as objects to be reformed and awakened? But in his more interesting moments, at least, his homelessness amongst the *kisans* pursued a politics that sought to both reject the logic of transcendence, and to dramatically radicalize the politics of neighbourliness. Perhaps Yagnik suspected (like many others involved in radical politics, including his associate Swami Sahajanand) that the marginal and the subaltern could never be a home, that marginality and subalterneity involved a homelessness that yearned for a home. To produce an ethnography or history of everyday subaltern lives, to claim that such history and ethnography constituted their homes, and to claim residence in these homes, such acts would have been to be more faithful to history and ethnography than to the marginality of the subaltern. Barred by his politics from claiming a home, Yagnik became homeless. By being carefully attentive to the 'positive' modalities and registers of such homelessness, perhaps we can engage more seriously with the alternative worlds that subaltern politics may point to.

Acknowledgements

This article – begun for a conference in 1997 at the University of Virginia, Charlottesville in honour of Walter Hauser – is dedicated to the memory of Rosemary Hauser (1928–2001). In the long period of its making it has benefited from many conversations and for these I thank Arun Agrawal, Shahid Amin, Dipesh Chakrabarty, Sudhir Chandra, Partha Chatterjee, Alon Contino, David Hardiman, Walter Hauser, Qadri Ismail, Pradeep Jegannathan, Priya Kumar, Allan Megill, Parita Mukta, Gyan Prakash, Gloria Goodwin Raheja, Tridip Suhrud, and Babu Suthar.

I thank Babu Suthar for suggesting how this complex formulation should be translated.

Notes

1 Both here and in other extended quotations, I have depended greatly on the 1986 type-script translation of Yagnik's *Atnwkatha* by Devavrat Pathak, Howard Spodek and John R. Wood, deposited at the University of Pennsylvania Library.
2 Munshi and Yagnik were fairly close in the 1910s, and even founded and ran a journal, *Navjivan ane satya* (New Life and Truth), together. There was almost a contrapuntal element to the ways their personal and political careers intertwined and diverged. Munshi was a key figure in the nationalist movement, though far more part of the mainstream. Like Yagnik, he was trained to be a lawyer but, unlike Yagnik, he was extremely successful. Both were deeply committed to the idea of companionate marriages and regarded themselves as confined in their initial years to traditional marriages that did not meet their ideals. Both their wives died while still in their twenties: Munshi then married Leelavatiben, a reformist widow from Ahmedabad; Yagnik remained single. In the 1930s, as Yagnik, having left the Congress, was getting more deeply involved in *kisan* and trade union activities, Munshi joined the Congress government in Bombay. As Home Minister, he was at the forefront of efforts to put down strikes by trade unionists. In the 1940s, Munshi went on to be a key assistant to Sardar Vallabhbai Patel and directed the 'police' operations that led to Hyderabad becoming part of the Indian state despite the Nizam's opposition; around this time, Yagnik was involved in organizing the peasants in the princely states against landlords. In the years immediately after independence, Munshi went on to become the Food Minister in the central government and to play a key role in ensuring that Hindi was the official Indian language; Yagnik languished in his *ashram,* fretting at his political irrelevance. When the movement for a separate state of Gujarat started, Munshi remained aloof; Yagnik, in contrast, was the principal leader of the Mahagujarat agitation.
3 The novel was part of a trilogy which focused on the rule of the twelfth century Gujarati king Jaysimha Solanki. The first volume in the trilogy, *Patanni Prabhuta* (1916) is set in the time when Jaysimha was still very young; in *Gujarat no Nath,* Jaysimha is beginning

to assert himself; and in *Rajdhiraj* (1922), Jaysimha is near the height of his powers. In the quotations from the text, I depend on the translation by N.D. Jotwani: N.D. Jotwani (tr.), K.M. Munshi. *The Master of Gujarat: A Historical Novel,* Bombay, 1995.

4 For some of Munshi's remarks on history, see Munshi (1956, pp. 519-526).

5 This is a collection of a series of essays by Narmad written between the 1850s and 1870s.

6 *Atmakatha,* Vol. 6. Yagnik died in 1971, at which time he was still writing the sixth volume, and had reached only as far as the Mahagujarat movement. The volume is thus incomplete. The writer Dhanvant Oza, who prepared the volume for publication, has included in it a selection of Yagnik's writings from later years in order to indicate the range of issues that Yagnik was involved with.

7 The remarks in this paragraph and the next are based on the *Atmakatha,* Vol. 6 and also on Brahmakumar Bhatt, *Le ke rahenge Mahagujarat,* Ahmedabad, 1994.

8 See, for instance, his discussion of the marriage between Mansukhlal Master and Taraben Master in *Atmakatha.* Vol. I, p. 139f. Hansaben Mehta, Leelavati Munshi, Premleela Khandvala, and Sharadaben Mehta, among others, are presented in his *Atmakatha* as·other women capable of companionate marriage. See especially Vol. 6, Ch. 9 of his *Atmakatha.*

9 Yagnik discusses his marriage at length in *Atmakatha,* Vol. 2, Ch. 13, partially in response to those who had criticized the first volume for not saying much about it. 'Why did a sincere worshipper of women like me commit this sinful deed'? Once he had married, he 'had no right to refuse it [the marriage] unilaterally'; he should at least have given her a place in his heart as a lifetime companion. 'This ... analysis is not my defence but a confession of my mistake, my secret sin. I hope the reader will accept it as such.

10 It is a different matter that Yagnik may not actually have been very celibate. According to David Hardiman:

> In Ahmedabad, it is often said that Indulal was distrusted because he committed the two greatest of sins for a 'Gujarati' (read: for a person having to operate within a Gujarati baniya culture), namely he was sexually loose (there were numerous rumours about his dallyings with the wives of prominent men) and also could not be trusted with money (he 'squandered' it, did not keep careful accounts. He lacked that careful calculating and moralistic approach to institution building which was so conspicuous a feature of the nationalist movement in Gujarat.
>
> (Personal communication, 14 July 1999)

Certainly, the allusions in his *Atmakarha* must have contributed to these rumours.

11 Yagnik's earlier efforts at helping *adivasis,* and the tensions that these caused with Vallabhbhai, are the subject of *Atmakatha,* Vol. 2, Ch. 12.

12 The tension between these two very different renderings of the popular is particularly forceful in the writings of Swami Sahajanand, a figure with whom Yagnik was closely allied. See, for instance, his *Mera Jivan Sangharsh* (My Life Struggle), published by his Shri Sitaram Ashram at Bihta in Patna district in 1952, two years after the Swami's death (translated by Walter Hauser and Kailash Jha as *My Life Struggle.* New Delhi: Manohar, 2018). See also Hauser's 'Swami Sahajanand and the Politics of Social Reform. 1907–1950,' *Indian Historical Review,* Vol. XVIII (1–2), July 1991 and January 1992, pp. 59–75. Yagnik was a great admirer of Sahajanand and dedicated the fifth volume of his *Atmakatha* to the Swami. The dedication described the 'Danda swami' as a fighter against zamindari, and as a figure who had lavished unparalleled *prem* or love on the *kisans* or peasants. Yagnik and Sahajanand were key figures in the setting up of the All India Kisan Sabha, the premier organization that lead peasant struggles in the 1930s and 1940s.

13 For arguments regarding the *ashram,* see Skaria (2002) and Confino & Skaria (2002).

14 Personal communications from Babu Suthar: 16 July 2001 and 20 July 2001.

15 For a more extended discussion of these two modalities of neighbourliness and of Gandhi's practices of translation between Gujarati and English, see Skaria (2002).
16 'Swadeshi Vrat,' *Navjivan,* 31 May 1931; *Akshardeha,* Vol. 46, p. 266; *CWMG,* Vol. 52, p. 209; 'The Law of Swadeshi,' *Young India,* 18 June 1931 (translation modified).
17 I thank Qadri Ismail for pointing out to me the importance and stakes of this question of singularity.
18 The work done in these *ashrams* is described at length in *Atmakatha.* Vol. 3, Ch. 1.
19 Letter from Indulal Yagnik to Amrital Thakkar, published in *Navjivan,* 4 November 1923; reproduced in *Atmakatha,* Vol. 3, pp. 346–351; see also p. 317.
20 Letter from Indulal Yagnik to Amrital Thakkar, published in *Navjivan,* 4 November 1923; reproduced in *Atmakatha,* Vol. 3, pp. 346–351, p. 317.

References

Anderson, B. (1991). *Imagined communities: Reflections on the origins and spread of nationalism.* Verso.

Confino, A. and Skaria A. (2002). 'The Local Life of Nationhood', *National Identities,* Vol. 4:1.

Dave, N. (1975). *Narmagadhya.*

Gandhi, M. (1908). Sarvodaya-IX. In *CWMG* (Vol. 8). Government of India.

Gandhi, M. (1919). In *CWMG* (Vol. 18). Government of India.

Gandhi, M. (1919). Speech on Swadeshi. In *CWMG* (Vol. 18). Government of India.

Gandhi, M. (1919). The Duty of Satyagrahis. In *CWMG* (Vol. 18). Government of India.

Gandhi, M. (1919). The Swadeshi Vow. In *Collected works of Mahatma Gandhi (henceforth CWMG)* (Vol. 18). Government of India.

Gandhi, M. (1919). The Swadeshi Vow-I. In *CWMG* (Vol. 17). Government of India.

Gandhi, M. (1928). In *CWMG* (Vol. 42). Government of India.

Gandhi, M. (1932). *From Yeravda Mandir.*

Gandhi, M. (1944). Letter to A.N. Sharma. In *CWMG* (Vol. 85). Government of India.

Gandhi, M. (n.d.). A History of the Satyagraha Ashram. In *CWMG* (Vol. 56). Government of India.

Gandhi, M. (n.d.). In *CWMG* (Vol. 10). Government of India.

Gandhi, M. (n.d.). In *CWMG* (Vol. 56). Government of India.

Gandhi, M. (n.d.). In *CWMG* (Vol. 56). Government of India

Gandhi, M. (n.d.). Letter to Sahebji Maharaj. In *CWMG* (Vol. 66, p. 29). Government of India.

Gandhi, M. K. (1997). *Hind Swaraj and other writings* (A. Parel, Ed.). Cambridge.

Khilnani, S. (1997). *The idea of India.* Penguin

Munshi, K. M. (1952). *Master of Gujarat* (N. D. Jotwani, Trans.). Bharatiya Vidya Bhavan

Munshi, K. M. (1956). Sparks from a Governor's Anvil, *Bombay, Information Directorate, Uttar Pradesh, Lucknow*

Nehru, J. (1981). *The discovery of India.* Jawaharlal Nehru Memorial Fund.

Satyagrahaashramno Itihas. (n.d.). (Vol. 50). Akshardeha.

Sheth, J. (1979). *Munshi: Self-sculptor.*

Skaria, A. (2002). 'Gandhi's Politics: Nationalism and the Question of the Ashram'. *South Atlantic Quarterly* Vol. 101:4.

Skaria, A. (2003). The primitivism of development. In K. Sivaramakrishnan & A. Agrawal (Eds.), *Regional Modernities in South Asia.* Stanford University Press. pp.

Yagnik, I. (1955). *Atmakatha* (Vol. 1). Jivanvikas.

Yagnik, I. (1955). *Atmakatha* (Vol. 2). Gujaratma Na Jivan.

Yagnik, I. (1956). *Atmakatha* (Vol. 3). Karavas.

Yagnik, I. (1971). *Atmakatha* (Vol. 5). Kisankatha.

Yagnik, I. (1973). *Atmakatha* (D. Oza, Ed., Vol. 6). Chhela Vehan.

9 The Homeless Gandhi

Vinay Lal

Ghar Wapsi

Mohandas Gandhi had, it may be said, mastered the art of dying (Lal, 2014, pp. 329–336). He was not going to go out from cardiac arrest, a stroke, cancer, or lung disease. He had declared, more than once and as recently as a day or two before his death, that he would either live to 125 or die with the name of Rama on his lips. Gandhi would be shot dead on the evening of 30 January 1948, and it is said that as the three bullets from a gun fired by Nathuram Godse felled him, he died uttering the words, 'Hey Ram'. His assassin, Nathuram Godse, a Chitpavan Brahmin who nursed a number of grudges against the man anointed as the 'Father of the Nation', disputed that Gandhi had said any such words; his brother, Gopal Godse, writing on this years later, agreed – and, perhaps in an effort to ensure that no dignity should be attached to that supreme moment, added that Gandhi had merely 'grunted' (Lal, 2001, pp. 34–38). Perhaps some others, who may not be thought to hold the predictably partisan views that we associate with the Godse brothers, but who nonetheless do think that Gandhi was a person of authoritarian disposition, might aver that the Mahatma sought, to use a contemporary expression, to 'control the narrative' around his death. But, if he did, it was not entirely with success. Admittedly, most people in India mourned; some cheered. More than a few, most pointedly, held him chiefly responsible for the vivisection of India and declared that he, more than Muhammad Iqbal or Muhammad Ali Jinnah, had played a critical role in birthing Pakistan; in the days before his death, Gandhi had been taunted by some as 'The Father of Pakistan'. Nathuram Godse ensured, in his defence, that this taunt and canard would stick; as he wrote in justification of his assassination of Gandhi,

> He has proved to be the Father of Pakistan. It was for this reason alone that as a dutiful son of Mother India I thought it my duty to put an end to the life of the so-called Father of the Nation, who had played a very prominent part in bringing about the vivisection of the country—our Motherland.
>
> (Godse, 2003, p. 113)

There is much that is interesting here, not least being the enactment of the Oedipal plot, but we cannot be detained by such considerations. Nathuram Godse held that

DOI: 10.4324/9781003544524-14

Gandhi was an effete man whose womanly ways and petulant behaviour, which led the old man to fast whenever he could not get his way, had emasculated the country. He had enough of what he took to be nonsense: spinning, fasting, listening to the inner voice, indulging women and reveling in their company, and departing from the rules of the world of muscular politics at one's will. Women, and not only his wife Kasturba, seemed to be everywhere around Gandhi: his grand-nieces, Abha and Manu were likened to his walking sticks; his personal physician was none other than a woman, Dr Sushila Nayar; Mirabehn, an English woman from an aristocratic family who had abandoned home and hearth to live by Gandhi's side since the early 1920s as his personal aide, was always hovering over him; a score of other women, including Sarojini Naidu and Rajkumari Amrit Kaur, as well as women with whom he nurtured what might be described as platonic relation-ships, constituted part of his inner circle. To what extent all this may have bothered Nathuram Godse, who was sworn to celibacy – as was, of course, Gandhi himself – is a matter of speculation, but the salience of these observations may be under-scored thus: in the rough-and-tumble world of politics, a world dominated by men and in which the Mahatma had himself been a key world player for some decades, Gandhi seemed to be most at home in the company of women. An extensive com-mentary on Godse and the politics of masculinity is scarcely necessary here, and Ashis Nandy's four-decade-old foray into this matter, '"Final Encounter": On the Politics of the Assassination of Gandhi', remains the most compelling interpreta-tion of the murder of Gandhi and the sources of Chitpavan Brahmin anxiety about Gandhi (Nandy, 1980, pp. 70–98). The previous attempts on his life, barring one just a few days ago before January 30 by the young Punjabi refugee, Madanlal Pahwa, had all been by Chitpavan Brahmins, and it is surely not accidental that the explicitly militant response to colonialism, which celebrated the real or alleged martial past of India supposedly obscured by colonial historiography, was steered by Chitpavan Brahmins.[1] The community as a whole, Nandy has argued, felt espe-cially aggrieved at the loss of its power as a consequence of British rule, and having a Gujarati bania such as Gandhi at the helm of power was not calculated to make Chitpavans, who bemoaned the loss of their masculinity, feel emboldened as the sun began to set on British rule in India.

Thus, in Godse's view, Gandhi deserved to die. Those who have come to adopt his view, especially in the last decade, are legion. Some who are entrenched in power make a show, as they must, of garlanding his statues and mouthing the cus-tomary platitudes about Gandhi's 'continuing relevance', but privately they con-sider Godse rather than Gandhi a martyr to the cause of the nation and the holy land that is India (*punyabhumi*). Godse, unlike most of them, was quite candid in hold-ing forth that India could never become a powerful nation-state that the rest of the world might envy so long as Gandhi was alive to guide the country's destinies. The assassin was also genuinely reverential in his feelings toward Gandhi, a part of his story that is little recognized: the Mahatma loved the nation and had awakened the slumbering masses, so Godse thought, but he had deviated from the path and gone astray (Godse, 1987). Gandhi, that inveterate user of trains, had derailed the coun-try. His murder would be the first step in the yet unnamed project of *ghar wapsi*:

even as Gandhi was being dispatched to his Maker, the country would supposedly be returned to its roots and the Muslim would be returned to the Hindu fold.

He Without a Home

Freud in *Civilization and Its Discontents* described the battle within everyone between *eros* (the instinct to love) and the death wish (*thanatos*). While we need not be beholden to Freud's precise reading of the death wish, it may be said that, in a peculiar way, Gandhi did not mind being killed. By this I do not merely mean what most who are familiar with Gandhi's life will at once infer, namely, that he often spoke in the last few years of his life, and particularly in the aftermath of the partition bloodbath, of having lost the desire to live. He had a premonition of his own death, and yet he had also said on more than one occasion, as I have already intimated, that he wished to live until he was 125 years old.

I have in mind something quite different, something in part, but only in part, brought home by the grating and increasingly oppressive culture of inhospitality that has gripped the Republic of India. It is grating because for all the critiques launched by liberals and the left of India's proud claims to 'tolerance', there is a case to be made for India as a land that has been hospitable to many cultures and the adherents of vastly different faiths (Gandhi, 2008). Dalits may well be within their rights to pronounce the idea that India has ever been 'hospitable' as preposterous, and Ambedkar himself was certainly of the view that the very notion of equality was utter anathema to most Indians and doubtless to upper-caste Hindus. One generally tolerates others, it may be said, from an assumed position of superiority, much in the way in which white Americans have 'tolerated' non-white immigrants – though it is yet a different matter how white Americans, immigrants as much as others whom they tolerate, came to assume this prerogative. But tolerance and toleration do not exist in the same register, and neither do 'hospitality' (which assumes some notion of the ecumene) and 'tolerance', and it is still another argument that an offer of hospitality need not rest on the presumption of unconditional equality. Moreover, even those who sneer at the idea of India's claim to tolerance would surely have to admit that the generation or two of Gandhi, Nehru, Tagore, Kamaladevi Chattopadhyay, and Maulana Azad had a capacious and ecumenical conception of India and that they displayed a remarkable catholicity in their embrace of ideas they found compelling without much heed to their origins. What I have called the culture of inhospitality has replaced such ecumenism and its tentacles have spread over large swathes of middle-class India in recent years, and I suspect that many in the generation that wrought India's freedom would have found India strangely foreign.

Though there have been many compelling interpretations of his life, Gandhi has increasingly struck me as someone who felt himself at sea in the world. Everyone has her or his own Gandhi: political activists, nudists, vegetarians, environmentalists, prohibitionists, civil resisters, celibates, anti-abortionists, and pacifists are only some among the many constituencies that have claimed him as their own and sometimes even adopted him as their mascot. It is time for the *homeless* to claim him as their own, though we should first strive to unravel a few of the meanings of home and dispossession. Conventionally, a man and a woman together made a home – though, in most cultures, the act of making a home has been deemed incomplete without children. Sometimes, in making a home, we may be dispossessing others by our act. When the home that we longed for is realized, it may lose its attractions. That home which was a refuge from the vagaries of human existence might come to burden or haunt us, creating other forms of dispossession. We may be more at home in something other than the home in which we live.[2]

We may seek to guard some in our home with a Lakshman rekha, keeping them in and keeping others out, and thereby help shape conceptions of the outside and the inside, the other and the self, the alien and the familiar. We may find that our home suffocates us while others experience the confinement as liberating; some seek to escape home and some seek to return home. Some find themselves unable to leave home; others find it insufferable to return home. One's home may be the repository of the sweetest memories, but it may also call to mind intensely troubling memories of abuse. Some return home and find it a death-trap; others return home and find themselves unrecognized. Agamemnon, king of Mycenae, returned home ten years after the onset of the Trojan war only to have his Queen, Clytemnestra, plunge a dagger into him as he relaxed in the bathtub; Odysseus's homecoming was less calamitous, though only his aged and faithful dog, Argus, could recognize him at once. The writer may be more at home in writing than in the home where he does the writing; some have even been known to make the jail a home. We may, like the reluctant exile, gain a political home and lose our cultural home. Oddly enough, we may have several homes, and yet feel dispossessed, or we may have no home at all and feel, if not the world at our fingertips, entirely liberated. We may be at home in not being at home at all, and the home that we call home may have no relation to the home that is in the heart.

Homeless in India – and Beyond

Was Gandhi ever at home? His life offers fleeting impressions of someone who, even as his feet were firmly planted on the ground, was curiously unmoored. For much the greater part of his adult life, Gandhi was bereft of a family home; sharing not even an extended family home that was overwhelmingly the norm in his lifetime. He shared his life not merely with Kasturba and their sons but with dozens and often hundreds of inmates in communes and ashrams, and was deeply resented by some members of his family for being insufficiently attentive to them and their needs. If, for instance, the notion of home implies the idea of a private sphere, Gandhi displayed not merely indifference to the idea of privacy but was inclined to see it as a species of secrecy and thus, deception. It cannot be an accident that, having vowed not to return to Sabarmati Ashram until India had been delivered from the shackles of colonial rule, Gandhi set out on the Dandi March in March 1930 and then seemingly drifted around, somewhat like a homeless man, for a few years until he settled upon Wardha in central India. Few have asked where he went after the completion of the Dandi March in April 1930, and where he was lodged when he was not in one jail or another, until he secured another place where he could establish another ashram. In early April 1936, he set himself up in the desperately poor and mosquito-infested village of Segaon, which then had a population of less than 700. Segaon had the virtue only of being, it is said, the dead centre of India, a comparatively barren spot that in its very emptiness could alone encompass the plentitude of India. Yet, it is questionable whether one can speak of him as 'drifting' around: wherever he went, the practices of the ashram went with him. The ashram was never only a physical site: the observances made it a portable unit, and both he and his followers could feel at home.

Gandhi was, nevertheless, beginning to feel homeless in the India that was taking shape even before partition tore apart his country and his heart alike. It is a long story, presaged in his first public pronouncements and appearances after his return to India following a 20-year stint in South Africa. The country he had returned to was something of a foreign land. He felt alienated not only by the uncleanliness in the streets and the filthiness of Hindu temples but, as he made clear in his speech at the inauguration of the Benares Hindu University, the reluctance, if not inability, of educated Indians to speak in their mother tongue. Over the course of the next two decades, he crisscrossed the country, often venturing forth into villages where scarcely anyone had ever cast a foot. It is doubtful that anyone in his generation traveled as much as he did; as he moved around, the country became at once both familiar and remote – familiar because the masses adopted him as one of their own, and he took on their pain, suffering, and grievances as his own; remote because he could not but feel how much his own thinking was at odds with the thinking of those who had declared themselves his acolytes and followers.

Hind Swaraj, his treatise of 1909, is but the most palpable evidence of the fact that Gandhi was an early critic of what post-World War II would begin to be called 'development'. This would put him at odds with many who would become the architects of a new India in the wake of independence. But he was also, and this is the greater irony, in view of his role as the principal contributor to the Indian independence struggle, never at home with the idea of the nation-state – an accursed entity, one parasitic on the idea of the 'foreigner' and calculated to produce obedient subjects, encourage conformity, reward homogeneity, secure the supposed integrity of borders, and demand and reward a juvenile patriotism. No nationalist was less invested in the nation-state that he had helped to forge – and, remarkably, once the nation-state had been achieved, he called for the dissolution of the Indian National Congress, drafting a statement in the hours before his death, advising each of its some million members to take up a village in the vast oceanic circle of villages that is India as their home (Gandhi, 2001, pp. 333–335). That is one of the measures of his greatness and of his distinct mode of being (at home) in the world.

Meeting with Ambedkar for the first time on 14 August 1931, Gandhi appealed to him to put aside their differences and work with him in the interest of the country. It is reported that Ambedkar famously replied, 'Gandhiji, I have no homeland'. The exchange is often reproduced as supposedly incontrovertible evidence of the divide between the two and the privileged position from which Gandhi approached the better educated Dalit leader who pointedly told him, 'How can I call this land my own homeland and this religion my own, wherein we are treated worse than cats and dogs, wherein we cannot get water to drink?'[3] Little could Ambedkar have known that Gandhi, who Ambedkar implied did not have to worry about a homeland, would just become his statues, though those are not being spared either these days. I suppose Gandhi, the putative 'Father of the Nation' who is increasingly being eviscerated from his 'homeland', would not mind the company of Ambedkar in being bereft of a 'homeland'.

One might reasonably find consolation in the words of Hugo of St. Victor, a twelfth-century scholastic theologian who wrote,

> The man who finds his homeland sweet is still a tender beginner; he to whom every soul is as his native one is already strong; but he is perfect to whom the entire world is as a foreign land. The tender soul has fixed his love on one spot in the world; the strong man has extended his love to all places; the perfect has extinguished his.

Yet, here again, there is reason to pause, if only to consider how the words of this Saxon monk have been bequeathed to us. Edward Said describes these lines as 'hauntingly beautiful' and seems to have been very much moved by Hugo of St. Victor: they surface repeatedly in his writings (Said, 2000, p. 185).[4] Perhaps Said found these lines 'hauntingly beautiful' because, awed as he was by the thought that the 'perfect' soul has extinguished his or her love for any one place, for that place we can call home, it was not given to him, the Palestinian intellectual who to the end of his life remained tethered to the idea of exile, to do the same. Moreover, Said, for all his intellectualism and the capacious humanism for which he has been celebrated, was as much an aesthete as anything else. It may even be argued that he was ultimately at home in the world of the aesthete as Gandhi was in the world of *ahimsa* – with the difference that in being at home in *ahimsa*, Gandhi found himself at odds with nearly everyone else. Where those who declared themselves his disciples and admirers saw nonviolence merely as an instrument to some end, to be deployed or forsaken as a matter of policy, for Gandhi, nonviolence was a matter of creed, a way of being in the world. There was only the very rare kinsman he might encounter in this quest: 'If a man does not keep pace with his companions, perhaps it is because he hears a different drummer. Let him step to the music which he hears, however measured or far away' (Thoreau, 1970, p. 442).

I am haunted, indeed everyone must be haunted, by the thought that the likes of Gandhi must perforce always be homeless in the world.

[This is a considerably enhanced, revised, and much lengthier version of an opinion piece first published in the *Indian Express* on 30 January 2018 and published online as well under the same title.]

Notes

1 In western India, a history of the modern militant response to colonialism might well begin with the founding of the Chitrashala Press by Vishnu Krishna Chiplunkar. One

of the earliest prints that was issued by the Chitrashala Press was of Nana Phadnavis, the renowned Peshwa minister. This was in 1884; five years earlier, Vasudeo Balvant Phadke led raids in the Poona and Satara districts against moneylenders and 'collaborators' of the colonial regime, leading the Governor of Bombay, Sir Richard Temple, to give it as his opinion to Viceroy Lytton that just:

> as the Mahrattas under Brahminial guidance once beat the Mahomedan conquerors bit by bit, … so the Chitpavan imagine that someday more or less remote, the British shall be made to retire into that darkness where the Moghuls have retired.

The audacity of the Chitpavans may be gauged by the bold assassination of the English official William C. Rand, Plague commissioner in Poona, carried out by the Chapekar brothers. They had the adulation of Bal Gangadhar Tilak, founder of the English-language *Mahratta* and Marathi-language *Kesari*, newspapers that the British took to be examples of the vile seditionist literature emanating from a disgruntled people. Tilak would in time go on to occupy a commanding place in nationalist politics. This is far from being a complete roster of the Chitpavan Brahmins who created a fertile ground for the spread of militant nationalism and claimed the Mahratta leader, Shivaji Bhonsle, as the source of their inspiration. See, for a more detailed analysis, Lal (2022); the quote from Temple is drawn from Tucker (1969, p. 347).

2 The last portion of this paragraph and the following are drawn, in a modified form, from my article Lal (2013).
3 The conversation does not appear in *The Collected Works of Mahatma Gandhi*. It is reported in Keer (1971, pp. 164-167), and most commentators and scholars who reproduce the conversation draw on Keer. It may well be the case that the conversation transpired as reported, but oddly enough the fiercely critical faculties that some scholars being to bear on their inquiries into Gandhi's disposition are put at abeyance in this case – all the more inexplicable in that Keer wrote a fawning biography, call it a hagiography – of Vinayak Savarkar, whose vehement hostility to Gandhi requires no elaboration. See, as instance, Menon (2022).
4 Said almost certainly first encountered these lines in an article by Erich Auerbach, whom he greatly admired and whose 'Philology and "Weltliteratur"'' [1952] he co-translated with Marie Said in *The Centennial Review* 13, no. 1 (Winter 1969), 1–17.

References

Gandhi, M. K. (2001). Draft constitution of congress. In *The collected works of Mahatma Gandhi* (2nd ed., Vol. 98, pp. 333–35). Publications Division, Government of India.
Gandhi, M. K. (2008). *Third class in Indian Railways*. Gandhi Publications League.
Godse, N. (1987). *May it please your honour* (3rd ed.). Surya-Prakashan.
Keer, D. (1971). *Dr. Ambedkar: Life and Mission*. Bombay: Popular Prakashan, [1954], 164–167.
Lal, V. (2001). "He Ram": The politics of Gandhi's last words. *Humanscape*, 8(1), 34–38.
Lal, V. (2013). "Gandhi and Palestine", *Critical Muslim*, no. 6: Reclaiming Al-Andalus, 171–190.
Lal, V. (2014). On the art of dying: Death and the specter of Gandhi. In J. Helfenstein & J. N. Newland (Eds.), *Experiments with truth: Gandhi and images of nonviolence* (pp. 329–336). Yale University Press and The Menil Collection.
Menon, D. (2022) "'What's in a name?: Those who invoke Ambedkar are complicit in a forgetting, much like Gandhi'", *Scroll.in (6 December 2017)*, online: https://scroll.in/article/859984/whats-in-a-name-those-who-invoke-ambedkar-are-complicit-in-a-forgetting-much-like-gandhi [accessed 12 November 2022].
Nandy, A. (1980). Final encounter: The politics of the assassination of Gandhi. In *At the edge of psychology: Essays on politics and culture* (pp. 70–98). Oxford University Press.

Said, E. (2000). *Reflections on exile and other writings*. Harvard University Press.
Thoreau, H. D. (1970 [1854]). Conclusion. In P. V. Doren Stern (Ed.), *Walden; Or, life in the woods* (p. 442). Clarkson N. Potter, Inc.
Tucker R. P. (1969). "The Proper Limits of Agitation: The Crisis of 1879--80 in Bombay Presidency", *Journal of Asian Studies 28,* no. 2, 339--355.

Index

A
adhvan 18, 19
adivasis 153, 169n11
Agni
    Agni Vaiśvānara 31
    Āhitāgni 30, 56
Agnihotra 35–42, 48–51, 54–56, 60, 65,
    66, 68n8, 69n32, 70n35
    Agnihotra cosmologies 51, 54
    Agnihotra offering 34, 39
    Agnihotra ritual 30, 48
    Agnihotrin 35–42, 47, 49–54,
        56, 66–68
    kāmya Agnihotra 56
    travel in the Agnihotra 30–32, 35,
        37, 40, 42
Agnyādhāna 32, 48
Agnyupasthāna 54, 57, 65
āhavanīya 30, 32, 35
ahimsa 158–160, 176; *see also*
    nonviolence
ahl-i-qalam 82
Ahmedabad ix, 146, 153, 158, 164,
    168, 169n7, 169n10
Akaniṣṭha 21
akhlāqi maṣnawī 84
Alagaddūpamasutta 52
Ambedkar, B. R. 173, 175, 176
    meeting with Gandhi 175, 176
Amīr Shihābuddīn Ḥakim Kirmānī 85
amūrta (unembodied) 42
Anandamath 112, 118
Anandniketan 163
Anderson, Benedict 147
Anhil, Bharvad 156
aniketa 163, 167
annalistic form of history writing 90
apratiṣṭhāna 20
apūrva 58, 59
aqueous
    aqueous aesthetic 120
    aqueous geography 112
    aqueous journeys 109, 124
    aqueous passages 110
Aravalli 113
Arhats 17
arthavāda 30, 49, 58, 59, 63, 65, 66, 68
asaṃjānānā 34
asceticism 31
    ascetics 44, 93, 96, 115, 118, 137
ashram
    grahastha-ashram 163
    Rashtriya Bhil Ashram 154
    Sabarmati Ashram 166, 174
    Satyagraha *ashram* 158
    Wardha and Sevagram *ashram* 174
Assam 97, 100, 163
asura 42, 54, 63, 120, 129, 135, 137
ātmabhāva 51, 57, 67
atmán 31–32, 41–47, 50–54, 57–59, 66,
    67, 68n12, 70n46, 70n50
    anātman 51, 53, 57, 67
    bhūtātman 45
    embodied *(dehin) ātman* 67
    jīvātman 45, 67
    paramātman 45, 67
ātmayājin 32
ātmayajña 32
attabhāva 51
Auerbach, Erich 177
Aufrecht, Theodor 61, 62
Avalokiteśvara 22, 54, 57
awakened mind 54
Awakened One 49
Awfi, Muhammad 80
āyatana 27, 52, 53, 57
Azad, Maulana 173

B
baaze jamin (badlands) 118
baḍāhākim 96, 102
Bahamanīs of the Deccan 89

Bahri, Malik 87–89
Barani, Ziauddin 86
Bariya-Kshatriya community 164
Bayly, C.A. 93
Benares Hindu University 175
Bengal 83, 84, 110–114, 117, 120,
    121, 158
    Bay of Bengal 109, 113–115
    Bengali literature 109, 110, 125n4
    Bengal Presidency 112
    Permanent Settlement of Bengal 117
    standard Bengali 124
Bhagavadgītā 54–56, 59, 66–68
Bhai Behen ke Geet 129
Bhairav rosary 133
Bhānujīdīkṣita 18
bhanwariyan 134
Bharatiya Vidya Bhavan 150
Bharat Mata 146, 147
Bharucha, Rustom 128
Bhils 145, 152, 154–156, 163, 164
Bhutan 71n66, 97, 100
bibliomania 111
Bikaner 140n5, 140n8
body
    corporeal body 44
    immortal body 32, 47, 48, 50, 55
    medieval body politic 79
    undying body 32, 41, 43, 52, 67
Bohlen, P.V. 63
Böhtlingk, Otto 61, 64
border 86, 89, 90, 92n4, 93, 100,
    103–105, 122, 175
    borderland 93, 94, 97, 98, 103
    misplaced aspiration 137
    trans-border individuals 8, 93
    trans-border religions 94, 105
Borunda 128
brahmacharya 150; *see also* celibacy
brahman 30–31, 42–45, 50, 55–57, 65,
    93, 95, 96, 102, 140n13, 149
    brahmanirvāṇa 55
    dashora Brahmans 86
    embodied *bráhman* 42
    parama-brahman 45
    unembodied *bráhman* 42
Brāhmaṇa 30–68, 70n52, 71n59, 71n71
    agnihotrabrāhmaṇa 30, 34, 42, 45,
        47, 50–52, 54, 55, 58, 59, 65–68
    Aitareya Brāhmaṇa 36, 62, 64
    apauruṣeya Brāhmaṇa 48, 59, 66
    Brāhmaṇa Padmanāba 60
    Brāhmaṇa texts 48–68
    Haug, Martin 62

Heesterman, J.C. 65
Houben, Jan E.M. 65
Jaiminīya Brāhmaṇa 32, 34, 36–39,
    41, 43, 48, 52, 65, 66
Kāṇva Śatapatha Brāhmaṇa 32, 37,
    46, 65–66
Mādhyandina Śatapatha Brāhmaṇa
    32, 39, 62, 66
Pañcaviṃśa Brāhmaṇa 36, 65
philosophy of the *Brāhmaṇas* 45, 49,
    51, 52, 58–60, 65, 66
Yajurvedic Brāhmaṇa texts 48
brahmand 161
Brahmasamāj 61
brahmin 31, 47, 49, 50, 52, 55, 60, 66,
    69n35, 150
    brahmin pandits 60, 66
    Chitpavan Brahmins 171,
        172, 177n1
British Army 98
Buddha, Gotama 49
    bhikṣus 18
    bodhi 54, 71
    Bodhisattvas 15, 17, 20, 22, 25n10
    Buddhadharma 51, 55
    Buddhadhātu 51, 57
    Buddha Dīpaṃkara 18
    Buddhahood 17, 21, 22
    Buddhas' travels 21
    Buddha Śākyamuni 18
    Buddhist masters 21
    Buddhist practice 20–21
    Buddhist thought 15, 23, 24, 26
    Buddhist travel 15, 22–24
    Complete Buddha 18, 25n13
    Mahāyāna texts 18
    Mahāyāna traditions 21
    mokṣa 58, 72n75
    nimitta 22, 27n42
    nirvāṇa 20, 21
    Parinirvāṇa 21
    Prajñāpāramitā 20, 26n25, 26n26
    Sanskrit Buddhist travellers 23
Buriganga 121
Burnouf, Émil 64
Burnouf, Eugène 64

C
Caland, W. 36, 65
Calcutta 61, 112, 119, 123, 125n8
caste 96, 97, 101, 102, 121, 129, 137,
    150, 164, 173
    caste system 96

celibacy 153, 154, 158, 163, 172; *see also brahmacharya*
census 94, 105n5, 148
chakkars 127; *see also* cyclical routes
Chakra 139
Chanda, Rani 109, 111, 112, 119, 125n2, 125n14
Chapekar brothers 177n1
Chattopadhyay, Bankimchandra 8, 109, 111
Chattopadhyay, Kamaladevi 173
Chavda, Vanraj 156
Chemjong, I.S. 102, 103
Chiplunkar, Vishnu Krishna 176n1
Chitrashala Press 176n1
Chōla 88
    Chōla kingdom 88
    Chōla rulers 88
Chukchinamba 98, 100, 101
circle 127, 139, 140n1, 145, 172, 175
circular routes 127
coastline 8, 109, 110, 112–114
Cohen, Signe 45
Colebrooke, Henry Thomas 60, 61
colonialism 60, 110, 111, 113, 120, 124, 172, 176n1
    'collaborators' of the colonial regime 177
    colonial episteme 5, 6, 8, 63; *see also* decolonial
    colonial historiography 172
    colonial history 24
    colonial limitations 24
    colonial rule 174
    postcolonialism 8
    pre-colonial Buddhist travel 24
    pre-colonial Buddhist world 24
conceptual geography 24
contact zone 93
cosmology 20, 31, 42, 54, 57
cyclical routes 127; *see also chakkars*

D
dalits 173
Dandi March 174
daridranayaran 155
Darjeeling 93, 94, 96–98, 100, 102–104
Das, Gyan Dil 93, 95–98, 104
Das, Narayan 131
Das, Shashidhar 95, 98
Daulatabad 82
Dawoa 122, 123
Debi Chaudhurani 112

decolonial
    decolonial episteme 6
    decoloniality 3
    decolonial mobility 4
    decolonial thinking 4, 8
    decolonization 15, 23, 24, 111, 113
Dehlavi, Hajib-i-Khairat 82
Dehlawī, Qāzī Badruddīn Dhārvāl 90
Deleuze, G. 30
Desai, Ramanlal 145, 151
desire 19, 30, 31, 43, 48, 52, 54, 56–58, 91, 113, 119, 120, 127–133, 135–137, 153, 160, 167, 173
    circle of desire 135
    desire line 135
    female desire 127, 133
    trajectory of desire 133
    transgressive female desire 133
Detha, Vijaydan 128, 129, 136, 140n9
Deussen, Paul 61
devāḥ/deva 32, 42, 45, 46, 53–55, 57
devatās 31
Dhaka 112, 119, 121, 123
Dhaleswari 121
Dhankuta district 96, 102
Dharma 8, 15, 18, 20–22, 43, 54, 58, 59, 93, 94, 96, 101, 163
    Abhidharma 19
    adhigama-dharma 18
    āgama-dharma 18
    dharmabhāṇakas 18
    Dharmakāya ("Dharma-Body") 18, 25n13, 51
    Kirat Dharma 94
    Satyahangma Dharma 93
    Wheel of Dharma 21
Dhār-Mandu 81, 84, 87, 90
Dhola-Maru 135
digvijaya 88, 89
displacement 125n11, 129, 130
diṭṭhiṭṭhānaṃ 52
Dodson, Michael 60
Drāviḍa kings 88
duḥkha 54
Dumont, Paul-Emile 65
Duperron, Anquetil 61
Durgeshnandini 112, 118

E
Eck, D. 6, 103
Eggeling, Julius 39, 40, 62, 66, 68n14
Eklingji temple 87
Englishtan 162

Enlightenment 4, 65, 66
eros and thanatos 173

F
fakir 145
Fārūqī, Ibrāhīm Qiwām-al-Dīn 84
Fenollosa, Ernest 16, 25n8
forest 21–23, 95, 111, 114, 115, 117,
    118, 125n12, 125n13, 147, 152; *see
    also* jangal
fort
    fort of Champaner 156
    fortress of Gangaratta 87
Foucault, Michel 6, 127
Four Truths of the Noble Ones 21
Freud, Sigmund 173

G
Gagraun 87, 89
Gandhi, Manubehn ("Manu") 172
Gandhi, Mohandas K
    alleged effeminacy of 172
    and the art of dying 171
    ashram life of 158, 159
    assassination of 171, 172
    encounter with Ambedkar 173,
        175, 176
    Father of Pakistan 171
    Father of the Nation 171, 176
    Gandhian politics 148, 158
    as homeless in India 174
    the Mahatma 171, 172
    murder of Gandhi 172
    national costume 5
    nonviolence 159, 176
    settles in Segaon (Sevagram) 175
garbha 34
gārhapatya 30, 34, 35, 69n23, 69n30
    gārhapatya fire 37, 39, 69n23
Geeton Ki Phulwari/Geeton ki Phulwadi
    128, 130, 135, 140n3
Geling 93, 97, 104
ghazi 71
Ghoomar 134
Ghosh, Aurobindo 65, 145, 153
ghumakkari 6–8; *see also* homelessness
Ghurid 80–81
Gītā 54–56, 59, 66–68, 163
Godse, Gopal 171
Godse, Nathuram 171, 172
Goldstücker, Theodor 64
Gonda, Jan 32, 33, 36, 48, 65, 68n13,
    69n26, 73n125

Governor of Bombay 177n1
Grassmann, Hermann 61
Griffiths, Ralph 61
Guattari, F. 30
Guhila lineage 87
Gujarat
    Gujaratan todadli 131
    Gujarati *bania* 172
    Gujarat no Nath 149, 168n3
    love for Gujarat 149
    Mahagujarat 151, 152, 154, 167,
        168n2, 169n6, 169n7
    Mahagujarat Parishad 152
Gurung 95–98, 100–102

H
hagiography 22, 177
hajj 79
Haribhadra 20
harijans 164
Hazrat-i-Dihli 80
Hegel, Georg Wilhelm Friedrich 4,
    62, 63
hijra 79
Hindu 3, 60, 63, 72n88, 92n2, 96, 97,
    101, 104, 113, 124, 135, 156, 164,
    173, 175
    filthiness of Hindu temples 175
    Hindu Bharat 150
    Hinduism 60, 72n88, 101, 150
    Hindustan 92n2, 111, 118, 125n6,
        150, 151, 162
historiography 79, 87, 172
home
    affinal home 129, 137
    conjugal homes 153
    cultural home 174
    family home 9, 174
    Ghar wapsi ("return home") 113,
        116, 171, 172, 174
    home and dispossession 174
    home goddesses 153
    homeland 100, 124, 131, 175, 176
    local home 147
    natal home 112, 135–138
    nationalist homes 153
    nation-home 147, 148, 151, 153, 154
    notion of home 174
    political home 174
    separation from home 91
homelessness 9, 145–148, 151–154,
    162–164, 167, 168, 173–176
    cosmopolitan homelessness 151

home and homelessness 9, 146,
   148, 162, 164, 167; *see also*
   ghumakkari
homeless in the world 176
homeless wanderer 163
Hooghly 112, 125n7
hostility to the nation-state 162
Housman, Alfred Edward 23
Hugo of St. Victor 176

I
Ichhamati 121
imagined communities 147,
   149, 159
   imagined cartography 131
   imagined circuit of mobility 127
   imagined maps 130
   imagined mobilities 129
India
   British India 62, 64, 93, 98, 100, 103
   capacious and ecumenical conception
      ofcentral India 9, 174
   dead centre of India 175
   East India Company 60–62, 66
   Indian art 63
   Indian intellectual history 30, 42,
      59, 68
   Indian theology 61
   as land of tolerance 173
   martial past of India 172
   middle-class India 153
   Northeast India 93
Indian National Congress 146, 154–
   157, 165, 168n2, 175
indigeneity 4, 9
   indigenous articles 159
   indigenous attire 5
   indigenous community 94
   indigenous groups 101, 140n10
   indigenous mobility 9n1
   indigenous tradition 94
Indology 30, 63, 65, 66; *see also*
   Orientalism
Indra 44, 132
injunctions 30, 49, 58; *see also vidhi*
Iqbal, Muhammad 171
Isami 82
Islam 80, 84, 87, 115
   centre of Islam 80
   dar-ul-Islam 80
   home of Islam 80
   Indo-Islamic polities 86
   Islamicate 79, 82, 83

itinerancy 103, 105, 140n9; *see also*
   homelessness
itinerant author-poet-performer 79
itinerant Sufi 79
religious itinerancy 103, 105

J
Jaisalmer 130
Jambudvīpa 18
Janaka 37, 38, 56
Janakachal 87
jangal 117
Jat 147
Jātakas 23, 26n23
jati 53, 95
Jaunpur 80, 83–85, 90, 92
Jinnah, Muhammad Ali 171
jñāna 15, 20, 44, 56, 57, 59, 66, 68
   ālayavijñāna 50
   avagati 15
   jñānayoga 56, 72n79
   puṇya and *jñāna* 15, 33, 48
   puṇyaloka 68n14, 33
   puṇyaṃ karma 32, 33
   vijñāna 43, 46, 50, 53, 57,
      70n41, 70n51
Jodhpur 128, 130, 140n5, 140n8
jogi 131–133, 135
Jones, Sir Williams 51, 60
journey 9n1, 15, 22, 32, 35, 39, 42,
   52, 80–85, 90, 92, 100, 109, 112,
   113, 115–117, 119–121, 124, 129,
   136–139; *see also* travel
joyous aesthetic 123
Jurewicz, Joanna 30, 32, 41, 45, 52, 65,
   68n12, 70n52, 71n59
Juzjani, Minhaj-us-Siraj 80, 81, 86
Juzjani, Qazi Minhaj-us-Siraj 80, 81, 86

K
Kalidasa 114
Kalimpong 96
Kalpi 80, 83, 85, 89, 92
Kapalik 115, 117
Kapalkundala 8, 109, 110, 112, 113,
   115, 116, 118
karma 32, 33, 40, 41, 43, 47–51, 53,
   55–57, 59, 66, 68, 71n72
   Karmakāṇḍa 56, 59
   kárman 33, 43, 45, 47, 48, 70n51,
      72n73, 72n78
   karmayoga 56
   karmic forces 21

karmic propensities 22
karmic retribution 47, 48, 50
meritorious karma 33
nityakarma 56
puṇyaṃ karma 32, 33
kavi Māhēśa 79, 86–90
    travels to Mewar and Malwa 79, 86,
        87, 89
Kaviraj, Sudipta 119, 125n4
kāya 51, 57
    Nirmāṇakāya 21, 51
    Sambhogakāya 51
    Saṃyuttanikāya 52
Keith, A.B. 33, 62, 65, 68n2
kelavni 153, 164
Khadaoda 87
    Khadaoda inscription 87–89
khadi 159
khal 121, 122
Khan, Qadr 90
Kheda 164
khedut 145, 156, 167
Khurasan 81, 90
Kisan Sabha 146, 169n12
Kokni 152
Kothari, Komal 128–131, 138,
    139, 140n3
Kṛṣṇa 54, 55, 59
Kuchhi oontni 131
Kuhn, Adalbert 64
Kumārila 57, 58

L
lākṣaṇika 16
landscape 9, 83, 91, 103, 109–114, 116,
    117, 119, 121–124, 135
    absence of landscape 109
    'desolate' landscape 109
    elevated location 109
    land as landscape 9, 110, 124
    landscape aesthetics 111, 117
    landscape of childhood 110
    landscape temporality 110
    literary landscape 111
    lost landscape 110
    past landscape 109, 110
    political landscape 83, 91
    religious landscape 103
    rural landscape 119, 122
    surroundings are revealed as
        landscape 109
Langois, Allexandre 61
Lefebvre, Henri 128, 139, 140n12

Lévi, Sylvain 64
liberal cosmopolitanism 151
Limbu 93–98, 100–104
    Limbu script 102
    Limbu villagers 93
    Limbuwan 100, 101
Lingden, Phalgunanda 93, 94, 97, 98,
    100–103, 105n3
linguistic community 152
loka 30–33, 35–37, 40, 41, 43, 44, 47,
    48, 50–53, 57, 65, 66, 69n26
    earthly *loka* 47
    puṇyaloka 33, 68n14
    svarga loka 33, 35, 36, 66,
        69n23, 69n24
Lorimer, Hayden 135, 140n10
Ludwig, Alfred 61

M
Macdonell, A.A. 64, 65
Magar (*jati*) 95, 101
Mahaguru 101–103
Maharashtra 151, 152
Mahratta 177n1
Maimadshahi turban 133
Maithili 111, 114
Malwa 79, 83–90, 92n3, 92n4, 130,
    149, 150
    Malwa court 83, 88
    Malwa-Mewar border 87
Mandalgarh 89
Mandasor 89
mārga 21, 22, 26n27
Marwar 128, 135, 140n11
masculinity, politics of 172
    loss of masculinity 172
    world of muscular politics 172
maṭh (monasteries) 97
medieval
    medieval courts 79
    medieval India 8, 79–81, 90, 91
    medieval literati 79
    medieval mobility 91; *see also*
        medieval travel
    medieval peregrinations 91
Medinipur 112, 125n7
Mehemedabad 164
Mehta, Munjal 149–151
Mewar 79, 86–89, 150
migration 9n1, 91, 110, 129
    bibliomigrancy 9, 109–111, 114, 124
    economic migrants 23
    emigration 79

immigrants 101, 173
literary migrancy 111
transmigration 60
Mill, James 62, 64
Mīmāṃsā philosophy 49, 56–61, 66–68
Mirabehn (Madeleine Slade) 172
missionaries 60, 66, 96
mobility 3–8, 9n1, 42, 59, 79–83, 85,
    88, 90, 91, 94, 95, 97, 104, 124,
    127–130, 133, 135, 137; *see also*
    sessility
    embedded mobilities 91
    gendered immobility 135
    history of mobility 137
    intensifying mobilities 91
    inter-sultanate mobility 85
    mobile narratives 137
    mobile wealth 137
    mobility and patronage 91
    mobility and sessility 91
    mobility in medieval India 79–81, 90
    movement and stillness 16, 25n6
    non-travel 24
    permissible mobility 133
    ritual mobilities 8
    spatial mobility 104
    trans-border mobilities 94; *see also*
        trans-border travel
    vernacular mobility 7, 95
    world of mobility and movement 80
Müller, F. Max 61, 64, 66, 72n94,
    73n124
Munshi, Kanaiyalal M. 145, 148–
    152, 155, 157, 168n2, 169n3,
    169n4, 169n8
Murshidabad 112
mūrta (embodied) 42

N
Nabakumar 113–116, 118
Nadiad 153, 159
Nagaland 100
Nāgānanda 23, 27n46
Naidu, Sarojini 172
; Nandy, Ashis 8, 10n2, 172
Narmad 150, 151, 169n5
nationalism i, ix, 5, 9, 103, 111, 113,
    120, 146–148, 151, 159
    anti-colonial movement 158
    anticolonial nationalism 111
    mainstream nationalism 146, 148,
        157, 158, 160
    national awareness 147

nationalist history 155
nationalist imaginary 147
nationalist paradigm of
    development 162
nationalist thought 147, 151, 155
nationalist tradition 153
nationalist violence 153
national luminary 103, 105n3
satyagraha movement 165
technologies of nationalism 148
nation-state 10n3, 24, 105, 110,
    161, 162, 165, 172, 175; *see also*
    imagined communities; hostility to
    the nation-state
Navjivan (Gandhi's main Gujarati
    weekly) 164, 166, 168n2, 170n15,
    170n19, 170n20
Nayar, Dr Sushila 172
Nehru-Patel-Munshi vision 157
neighbourliness 148, 159–165
    law of love 160
    logic of neighbourliness 159
    love for the neighbour 167
    padoshi 148, 159, 160
    padoshi dharma 160
    padoshi pahelo ('neighbours
        first') 160
    politics of neighbourliness 9, 146,
        148, 162, 164–167
    practice of neighbourliness 163
Nenpur 146, 164
Nepal viii, 93–105, 105n2, 105n5
néti néti doctrine 44
niketan 163, 164
Nimdihi, Abdul Karim 90
Nirukta 60, 64
nirukti 19
    method of *nirukti* 19
Nirvanadas 97
Nizami, Sadr al-Din Hasan 80, 82, 83
nomad
    mad nomads 3
    nomadic communities 129
    nomadic criminal tribe 5
    nomadic pastoralist 9, 131, 140n4
    nomadic societies 82
    nomad natives 5
    nomadology 3

O
Oertel, Hanns 62
Oldenberg, Hermann 61, 64
ontology 20

oral history 128
  oral performance 9, 131
  oral rhythm 128
  oral tradition viii, 4, 136
  oral voicing 136
Orientalism
  Orientalist 60, 62, 63, 66
  Oriental philology 61

P
Pahwa, Madanlal 172
Panchmahals 145, 154, 164
Panchthar 96–98, 100, 101, 104
paramparā 95, 105n4
Patan, Anhilwad 156
Patel, Vallabhbhai 151, 154–156, 163, 169n11
patriotism 147, 175
  desh prem (love of country) 151, 159
pauruṣeya 48, 66
Peehar va Sasural ke Geet 129
Persian 4, 60, 61, 80, 82–86, 91, 91n1, 110, 111, 118
  Persian literati 80–83, 91
  Persian New Year 84
  Persian vocabularies 85
Pinch, William 93
Poley, Louis 61
Potadar, K.R. 54
prajā 31–37, 39, 41, 42, 44, 47, 50–54, 56, 57
Prajāpati 30, 31, 33, 36, 41, 42, 44, 50–52, 57, 65, 69n19; *see also vāc*
prāṇāḥ 32, 34, 42, 45, 50, 53, 57, 68n8, 69n35
pratiṣṭhita 46, 55
pratītyasamutpāda 51, 53, 57, 71n59
Prem ke Geet 129, 136
punarmṛtyu 33, 35, 40, 41, 52–54, 68n15
Puṇṇaka 50
punyabhumi 172

R
Rabi 98, 104; *see also* Panchthar
Rajasthan 127–134, 137, 140n3, 140n11
  Ajmer 89, 130
  Chittod 130
  Chittorgarh 89
  Chittorgarh Kirtistambh 87
  Ker 131, 136
  Kotah 89

Kumbhalgarh 87, 89
Kumbhalgarh Fort 87
Medta 130
Rajasthani women 128–130, 135, 136, 139
Rajasthan Sahitya Akademi 134
Rajkumari Amrit Kaur 172
Rajput 79, 112, 137, 140n5, 149, 150
Rajputana 79
Rajput heroism 149
Rana (regime) 89, 93–96, 101, 103, 105
  Rana, Jang Bahadur 95, 96
  Rānā Kuṃbha 86–89
  Rana of Mewar 87–89
Rand, William C. 176n1, 177
Rangbul 97
Ranthambore 89
Rashtriya Swayamsevak Sangh (RSS) 153
Ratanakhera 87, 89
Raydhan 133, 134
Rebadis/Raikas 129, 140n4
Renou, Louis 65
rituals 30, 49, 55, 58, 65
  continuous ritual offering 31
  embryonic offering 35
  ritual activity 47
  ritual fires 32, 39
  ritual karma 56–57
  śrauta rituals 30, 49, 55
river journey 121
river passages 9, 110; *see also* river journey
Rosen, Friedrich 61
Roth, Rudolph 61, 64
Roy, Dwijendralal 149
Roy, Ram Mohun 61
ṛṣis 45
Rumjatar 96
rūpa 50, 53, 57
  nāma rūpa 31, 53
  surūpa 17

S
saat salaam 131
Śabara 58–59
safar 81; *see also* travel
Said, Edward 3, 7, 11, 16, 176, 177n4; and exile 129, 151, 173, 176
Said, Marie 177n4
salokatā 41
Sāman 32, 43
Sambhar 131, 136

saṁsāra 16–17, 19–21, 23, 49, 51, 53, 55–57, 59, 67
saṃsiddhi 56, 71–72n73
sáṁskaroti 32, 52, 53
saṁskṛta 19, 26n21, 26n22
samudra 51, 114
Śaṅkarācārya 45, 55–57, 71n72, 72n74, 72n76, 72n78, 72n79
sannyāsin 56
Sanskrit viii, 4, 8, 15, 18, 23, 25n1, 51, 60, 61, 64, 65, 86, 96, 110, 111, 113, 114, 119; Sanskrit College 72n94
Śāntideva 22, 27n41
Saraswati, Dayanand 145
śarīra 44, 51
Satyahangma 93–94, 97, 104–105
Saurashtra 149, 151
Savarkar, Vinayak 177n3
Sāyaṇa 30, 39, 45, 49, 61, 64, 69n20, 69n32, 70n48
Schlegel, A.W. 62
Schlegel, Friedrich 62
Schroeder, Leopold von 62
Schwab, Charles 60, 72n82, 72n85, 72n86, 72n98
sect 63, 93–95, 103, 105n4
    bhakti sect 3, 93, 95, 96, 103
    Josmani sect 94
sessility 8, 83, 91; *see also* mobility; preference for sessility 83
seva 159
Shadiābād 83–84
Shah, Prithvi Narayan 95
Shaiva 93
Shantiniketan 163
shapla 122–123
Shihab Ḥākim 79, 80, 83–86, 88, 90, 92n2, 92n5; *see also* Amīr Shihābuddīn Ḥākim Kirmānī
Shikoh, Dara 61
Shilgunisuri 156
Siddiqui, Iqtidar Husain 84
Sikkim 93–95, 97, 100, 102, 104
Silauti 101
Simmel, Georg 140n2
Singalila 93–94, 97–98, 103
Smith, Adam 62
Smith, Brian 32, 65
smṛtyupasthāna 54, 57
South Africa 158, 164, 175
Sovan moriyo 131
space 9, 10, 15, 20, 23, 86, 93, 109, 112, 116–118, 120, 124, 127, 133, 138–139, 139n1, 140n12, 148, 163, 166–167
    experiential space 139
    intermediate space 31, 51, 118
    Platonic space 127
    remaindered space 118
    social production of space 128
spatiality 111, 127, 128, 130
Staal, Frits 65
Stevenson, J. 61
stri-kelavni 153
Subachani 123
Subba, Jash Raj 94
sublimity 63
Sugatas 16
Sultan 80–89
    Delhi Sultans 86
    Gujarat Sultan 88, 90
    Lodī Sultans 89
    Malwa Sultan 80, 83–85, 87–89
    Sharqi Sultans 84
    Sharqi Sultans at Jaunpur 84
    Sultan Abu'l Muẓaffar Bārbak Shah 84–85
    Sultan Iltutmish 81
    Sultan Maḥmūd Shāh Khalji 83–84, 87–89
    Sultan Muhammad-bin Tughluq 81–82
    Sultan of Jaunpur 80, 83–86
    Sultan's military activities 89
    Sultan's quasi-military campaign 88
Sultanate 80–81, 83, 85–86, 88–90, 92
    Delhi Sultanate 80–81, 83
    Gujarat Sultanate 88, 90
    Malwa Sultanate 80, 83, 88
    medieval Indian sultanates 80
    Timurid Sultanate 90
Sunuwar 95
Sūrya 34, 69n18
Sūtra 15, 18, 20, 60
    Mahāyāna Sūtras 18
    pauruṣeya Śrautasūtras 48
    Śūraṅgama Sūtra 54
    Sūtrakāra 49
Suttanipāta 49, 52
suvargas 41
swadeshi 158–162, 170, 170n6
    swadeshabhiman 150
    votary of swadeshi 161
    vow of swadeshi 158, 160
swaraj 155, 158, 162, 170
    Hind Swaraj 158, 162, 170, 175
    swarajya 162

T
Ta'rīkh-i-Mahmūd Shāhī 88, 90
tabaqat 87
Tagore, Abanindranath 119
Tagore, Rabindranath 6, 112, 120, 123, 125n14
Tārā 22
tathāgatagarbha 51, 57
Tathagatas 17–18, 71n62
tawarikh 83, 87
Temple, Sir Richard 177
Teuscher, U. 87–89, 92
Thakkar, Amritlal V. 154
Thapa, Lakhan 95–96
Thar 9, 61, 131, 133, 137
Thite, G.U. 65, 68n1, 72n79
Thoreau, Henry David 176
Tibetan Tengyur 22
Tilak, Bal Gangadhar 177
Tonk-Toda 140n6
Trautmann, Thomas 60, 62, 72n87, 72n88, 73n104, 73n105
travel
    alternative travel 6, 8
    Battuta, Ibn 81
    leisure travel 6, 22–23
    medieval travel 79, 81
    ritual travel 30, 127, 133
    river travel 110
    roving authors 86, 90–91
    trans-border travel 8, 93–94
    travel and mobility 79–83
    travel for work 87
    traveling gaze 109, 125
    traveling theory 4, 7
    traveling writers 8, 80–81, 91
    travel narratives 9n1, 11, 109
    travel writers 80
    travel writing 3, 6
    vernacular travel 3–7, 9
    water travel 8, 109
travelogues ix, 6, 10–11, 110, 125–126
tri-adhva-samatikrānta 20
Tripathi, Kumud 153
Tull, Herman 47, 65
Tūnī, Abdul Ḥusain bin Hājī 90
Turnbull, Reverend Archibald 96

U
Udaipur 87, 140n5, 140n6
Umarkot 130
universalisation 128
    spiritual universalism 160
universal cosmic self 44
universal law 161
unsurpassable oceans 51; *see also* *samudra*
untouchability 154, 158, 165
upacāra 16
Upaniṣads
    Bṛhadāraṇyaka Upaniṣad 43–48, 55, 61
    Chāndogya Upaniṣad 44–45, 61
    Kaṭha Upaniṣad 44
    Kauṣītaki Upaniṣad 44
    Maitrī Upaniṣad 45, 70n46, 70n49
    Muṇḍaka Upaniṣad 44–45
    Śvetāśvatara Upaniṣad 44–45
    Taittirīya Upaniṣad 44, 70n44
Urry, John 127
urwatu'l-wuṣqá 84–85
Utkala 88
Uttara-Mīmāṃsā 57

V
vāc 30–31, 33, 36, 41–42, 44, 50–52, 57, 65, 69n19; *see also* Prajāpati
Vaishnava 93, 103
    Tamil Vaiṣṇava hymns 60
Vajrayoginī 22
Vaniya, Champa 156
Vārāṇasī 21
Veda 36, 46, 53, 57–59, 61, 64–65, 70n40
    Advaitavedānta 67, 68
    Benfey, Theodor 61, 64
    Bergaigne, Abel 64
    Bodewitz, Henk 32–34, 36, 39, 41, 43, 47, 48, 65, 68n3, 68n4, 68n5, 68n10, 68n15, 68n16, 69n18, 69n19, 69n22, 69n23, 69n29, 69n33, 69n35, 70n37, 70n38
    Ṛgveda viii, 40, 45, 60–62, 64, 69n30
    Rig-Veda-samhitâ 61
    Vedānta 56–57, 60–61, 66–67, 98
    Vedic ritual 30, 47, 54, 59, 67
    Yajurveda 61–62, 64
vernacular modernity 6, 10
Viceroy Lytton 177
vidhi 30, 49, 57–58; *see also* injunctions
village
    rural community 124
    rural dialect 112
    rural landscape 119, 122
    rural life 111
    rural ordinary 120

rural people 156–157
rural social groups 129
Virocana 44

W
walking 21, 98, 121, 127, 135, 140n10, 172; *see also* travel
wandering 4, 7, 82, 93, 97, 100, 102, 103, 127, 131–133, 135
aimless wanderer 131
wandering mendicant 3, 131
"legitimate" wandering figure 133
Wardha 158, 174
warrior king 88
waste 60, 117, 125n10, 125n11
wasteland 117, 118, 125n11; *see also jangal*
water 8, 43, 109, 110, 112, 114–116, 120–122, 127, 129–136, 140n4, 140n9, 176
backwaters 120
river water 136
water geographies 112
waterscapes 9, 110
waterways 112
Weber, Albrecht 61–62, 64
Weltliteratur 110, 177n4
Whitney, W.D. 61, 64
Wilkins, Sir Charles 60–61
Wilson, Horace Hayman 61–64

Windisch, E. 60, 64, 72n80, 72n80, 72n83, 72n85, 72n90–72n94, 72n98–72n102, 73n112–73n124
Witzel, Micheal 65
world
metaphorical world 24
worldliness 110–111, 116
world of ahimsa 176
world of longing and belonging 139
world of merit 33
world of the aesthete 176
World War II 175

Y
Yagnik, Indulal 5, 145–170
yajamāna 30, 32, 35, 39, 40, 48, 50, 53, 54, 68n10, 68n11, 69n32, 69n34, 71n57
yajña 42, 48, 54, 55, 59
Yājñavalkya 37, 43, 45–46, 66, 70n52
yajus 32, 43
yañña 49–50, 71n55, 71n57
yathābhūtam 53, 57
Yavana 89
Yeravada jail 164
yogācāra 50
Yumaism 94

Z
ziyara 79